OTHER PLACES AND OTHER WAYS TO FIND AND KEEP ARIZONA'S GOLD

Author testing behind boulders on Copper Creek, Yavapai County, Arizona, June 22, 1993.

Ronald S. Heilgus
Tucson, Arizona

Fig. A. **MUGGINS MOUNTAINS WILDERNESS AREA BOUNDARY**
Looking up Muggins Wash toward Muggins Peak in the distance, September 22, 1993.

YOURS FOR THE TAKING

OR

GETTING MORE OF YOUR SHARE OF DESERT

GOLD!

BY
RONALD S. WIELGUS

DEDICATION

I dedicate this book to the recreational prospector. May he find not only gold in his pan but the peace and satisfaction that comes from clean air, beautiful surroundings and honest work.

ISBN 0-9635601-1-5
YOURS FOR THE TAKING
OR GETTING MORE OF YOUR SHARE OF DESERT GOLD!
First Edition.

Cover Photo: Coarse gold nuggets, actual size, recovered by the author.

CONTENTS

Fig. B. **CAÑADA DEL ORO WASH - CAÑADA DEL ORO MINING DISTRICT**
Looking to Pusch Ridge and the Santa Catalina Mountains with Cañada Del Oro Wash in spring flow, April 1993.

Fig. C. **RUINS - KING TUT MINING DISTRICT, MOHAVE COUNTY, ARIZONA**
The remains of an adobe brick mine building, looking to the Southeast, August 18, 1993.

PREFACE

When I began writing ***Arizona's Golden Secret*** it was the culmination of over 30 years' experience prospecting for Arizona's gold. I thought back to those first days so long ago. A newcomer to Arizona then, like many before me I had heard only vaguely about people going out into the desert and mountains of this state to pan for gold. Gold, panning for gold, this was the stuff of dreams! Here I was, in a place that really had gold and all I had to do was go out with a pan and get it; but, where should I go and how do I get it?

Sadly, no one would, or could, tell me; nor did I ever see any of that gold purportedly found by gold hunters. From books I had read about the Fortyniners, they found gold in rivers and streams so all you had to do was pan the gravel and you, too, got gold. Well, that may have been true 150 years ago in the Mother Lode Country, but today?

Having aroused my curiosity I decided that I would go out and do some panning. I was living in Phoenix at the time and the only river I heard about was the Salt. Except in times of flood, the Salt River through the Phoenix area is mostly a dry river bed. Its flow below Stewart Mountain Dam is swift and deep and very intimidating. I ruled it out.

My next thought was the Verde where it flowed through the Fort McDowell Indian Reservation. So, with a friend also interested in gold panning, I tried my luck panning the gravel of the river bed. What we got for our efforts was much mica in our pans and the question, "Is this gold?" Well, after some discussion, we decided that it wasn't gold. Boy, were we dumb! Nevertheless, we had a heck of a good time and vowed that we would do it again only this time we would find gold. However, we never did go out again although we talked about it from time to time.

Finding gold was proving harder than I thought. There were two other guys at the place I worked, young like me, and eager to find gold. We discussed the possibilities endlessly. One of the guys, who came to Arizona as a baby with his parents, had heard that there was gold in Black Canyon Creek near Black Canyon City. In no time at all we made our plans and soon we were "prospecting" the river bed with shovels and pans. We didn't know a thing about gold prospecting, but we didn't know that.

The river bed was bone dry. The day was hot. We scouted the river bed, chose a spot and began digging. The three of us dug a hole in the sandy soil you could lie in. From the bottom of the hole we took out a few shovels full and panned this sandy dirt in our pans with our drinking water. We didn't find any gold. Hot and discouraged we called it quits and headed home.

One day at work an older guy came into our drafting room, a mechanical engineer who lived in Rock Springs, just south of Black Canyon City. He was consulting the architect I was working for at the time. We told him about our prospecting trip and how it was a bust. He just listened but offered no suggestions. Over the months, he came in regularly to the office for consultation and when through dropped by to chat with us. One day, when the conversation turned to gold, he brought out of his pocket a glass bottle. Large gold nuggets filled it to the neck. We were bug-eyed! This was the first real gold (meaning, natural gold, not jewelry gold) we had ever seen.

We bombarded him with questions. You know, like, "Where did you find them?" and, "How much are they worth?" Apparently, he was in an expansive mood and told us much about what we were yearning to know. He had been prospecting for many years and had gotten to know some real old-timers from whom he had learned the ropes for finding gold. The nuggets, he said, came from Soap Creek. He explained all about how he found them on the bedrock of a small tributary, in the cracks. We got a good lesson on gold prospecting that day along with his admonition that we not reveal the location where he found his gold. He feared that if the secret got out prospectors would overrun the place. This was in 1959. Back then, you still could see the old prospectors with their burros coming out of the Bradshaw Mountains, on occasions, when driving up to Prescott. Arizona in 1959 was still the "Old West," and Phoenix's population then was only 300,000. Tucson's was about 100,000.

Not long after our lesson, I went up to Soap Creek and followed the dirt trail to the spot the engineer had described in detail. It was here that I found my first nugget of gold and the thrill was indescribable. I guess that the gold bug bit me that very moment and I have been prospecting for Arizona's gold ever since. For more than five years, off and on, I made trips to the area. It was here that I learned the art of sniping and honed my eye for future prospecting.

There was a miner's cabin here at the time. Now long gone, it appears in a sketch I drew for my book ***Arizona's Golden Secret,*** on page 17, as it looked in 1963. A man named Gonzales lived here and mined the placer gold. It was from him that the engineer learned much of his craft. I never did meet Mr. Gonzales. It was late in 1965, after I made many prospecting trips without seeing any sign of the miner, that I got up enough courage to walk over to the cabin. You never know what to expect when you arrive unexpectedly at anyone's home, let alone a reclusive miner. No one was home. It appeared that no one had been living in the cabin for some time. Although everything the man owned was in place, his cups, coffee pot and the like, from the dust covering everything I could tell that he had gone. He had left everything just as it was and simply never returned.

After a hiatus of more than a decade, I decided to return to Soap Creek with my son, Roger, and a metal detector. We were going to the Gonzales cabin to see what we could find. The cabin was not there! After the passing of many years someone had torn it down and removed everything. The site, now overgrown with brush, was very obscure. Yet, we were able to find enough clues to pinpoint the exact location of the cabin. Thoughts of a cache of gold never left our minds as we methodically went over the site and surrounding area with the metal detector. Alas, if there ever was a cache it wasn't there, either. I told my son to continue detecting while I would look around for the dump, the place where the miner disposed of his trash. It took awhile but I found it down slope of the cabin site in a little ravine. Here was the accumulation of years of a man's life in the boonies. There were rusty cans by the hundreds, and broken bottles, but in all that pile were some real gems. Perfect liniment bottles revealed themselves as I probed through rusty cans, many still capped and some still containing a small amount of the liquid, now congealed. I uncovered some metal coffee cups, the kind with the blue and gray enamel, along with an odd assortment of discarded personal items. Finding the coffee cups was poignant for I remembered seeing some like these on my first visit to the cabin. It seemed that the man's whole life as a lone miner revealed itself to me in that dump, and it saddened me. His had been a frugal life, living alone, eating out of cans and rubbing the aches and pains from daily toil with strong liniment. We left, taking the few items I had found, the bottles and the coffee cups. We have never returned since.

Around the time of my first visit to Soap Creek I had heard about Lynx Creek and how it had been one of the biggest producers of placer gold in Arizona. I vowed that I would get to it and pan out some of its gold, and sure enough, I did just that. I continued to prospect other areas that I heard about through the grapevine, for then I was already a member of the "fraternity." I did most of my mining by sniping and a little with a home-made sluice box. I never heard of suction dredges then nor did I see any. They came much later.

I acquired in 1968 the one book that everyone looking for gold in Arizona should have, the "Bible," Bulletin Number 168 entitled ***Gold Placers and Placering in Arizona***. From this book, I learned about other placer areas but it wasn't easy finding those spots. The writer seemed purposely vague when it came to specifics, preferring to use the word "certain" many times over, for example: "Certain pockety channels..." and "Within the arcs of certain curves..." From my personal experience mining some of the areas mentioned, I began to doubt whether he really visited any. Much of his information was secondhand, hearsay and from prior reports, and so much of the book now is out of date. What's this about earning less than fifty cents a day? Even in the Great Depression of the 1930s this was very low pay. How did he arrive at these earnings figures? Much gold found then went unrecorded, not unlike today. I doubt that the miners then were any more truthful about what they made than those of today. I believe that they made much more. I only bring this out so that you won't be discouraged before you start. In spite of these criticisms I feel that the book's value as a source of information far outweighs its faults.

Well, I was mining gold at a time when its price was $35.00 an ounce at .999 fine and I wasn't getting rich, but I was learning and having a great time doing it. There were very few recreational prospectors then. This changed in the early 1970s when gold became a free market commodity. The sudden increase in the price of gold resulted in a proportionate interest in recreational mining. If you thought that miners were tightlipped when gold was $35.00 an ounce, you would swear they had their jaws wired shut when the price took off. Now, gold was worth fighting over. However, persistence wins out in the end and many new prospectors came on the scene. They, too, acquired lockjaw.

Now, after more than 30 years of prospecting for gold, I decided to share some of the secrets and wrote the book ***Arizona's Golden Secret***. This is the starter book for persons interested in finding Arizona's gold. The response from my readers has been gratifying! Encouraged by this response, I have written this companion piece ***Yours For The Taking***. The other places, and the other ways to get gold, are in this book. Like the first book, I have been there and I have done the mining in the ways I describe. I want you to have that knowledge and go out and get that gold! It's yours for the taking.

INTRODUCTION

I wrote this book in response to the many readers of my first book, *Arizona's Golden Secret,* who wanted more places to go to and other ways to mine gold. As some of the more popular gold placers become crowded with prospectors, many of you would like to reach other gold-producing fields that have a little more elbow room, particularly on weekends. These relatively remote or little-known placers do not have ground as heavily worked over as some of those popular, easily accessible areas.

Then, there are those just starting out in prospecting who would like to get gold but can't afford the necessary equipment just yet. Maybe, they would like to try a little gold prospecting before they make a commitment to buy anything but a gold pan. Others would like to mine gravel relatively untouched, or ignored, by other miners. Some of you want to know about using specialized mining equipment, such as high bankers, and mini sluices, that you have seen or heard about, before going ahead with another investment.

Prospecting for gold is somewhat like fishing. Every good fisherman has his favorite stream or lake and likes to go back to it often. He knows that his chances are good that he will get his catch, and if he is lucky, there is always the chance of catching a record. However, there are times when others find out about his good fortune and invade his spot, or perhaps he just longs for a chance to try another stream or lake. His adventurous spirit drives him to seek out new places, to make new discoveries. Chapter I is for the prospector who is like this fisherman. In it are directions to placer fields off the beaten path. A few are close to population centers but little known and see very few miners. Some are remote. These areas present many challenges to the prospector as most are lacking in water. They are good places to try out new ideas on dry recovery and a metal detector is very handy to have along. All have good gold. If you are confident in your skills as a prospector and want to take them on, they are ready and waiting for you!

For the prospector just starting out and on a limited budget, or the one not ready to go all out, my chapter on sniping is the answer. Not only will he get gold using a minimum of equipment, he will hone his prospecting skills, much of which is training the eye to read the terrain and see the things most others miss. Gold is where you find it but most often you find it in known gold areas than elsewhere. In these areas Mother Nature hides her gold well but she leaves little clues to where you can find it. You must be able to recognize those clues if you want to be successful at getting gold. In a way, sniping for gold is a form of testing the ground before you bring in the heavy guns but you can make it a way of getting gold all by itself. I started out in sniping for gold before I went on to more productive mining methods but I still do it today simply because it is a good way to get good gold, and I enjoy doing it. Chapter II tells how you can get gold by sniping.

If you have sufficient water available in the gold-bearing creek or stream, suction-dredging is one of the best ways to get gold. With a suction dredge of 3-inch size you can process many cubic yards of gravel in a day and the sheer volume of material processed is how you get much gold. That's why, in the Mother Lode Country of California where gold-bearing rivers and streams run full and deep, miners there can, and do, use dredges of 4-inch size and bigger. Not only that, a suction dredge is like a vacuum cleaner. It sucks up all the bedrock gravel, and when used properly is very effective in extracting almost all the gold. Many times, however, in the areas where you can use a suction dredge, miners rework the gravel over and over. This is particularly true for Arizona where gold-bearing creeks and streams with sufficiently available water are few.

Instead, in these localities, you can go for the gold most miners ignore or don't really know about; that is, the gold in the high banks. This is the gold left high and dry when the watercourse that carried the gold changed direction, or cut downward, to its present position. For thousands upon thousands of years, this gold accumulated, now locked in position in the gravel of the banks. Left high and dry, the banks erode more slowly when compared with stream bed gravel during heavy run-off. The gold in the high banks, for the most part, just sits there waiting for its turn to erode out sometime in the distant future. It is available to the miner with the right equipment, a high banker. If the miner with a high banker happens upon an ancient channel, he has the right equipment at hand to mine this bonanza. Chapter III discusses what you need to get this gold and how to get it.

A device gaining widespread popularity in Arizona is the mini sluice, also called a micro sluice. This little machine is a baby brother to a high banker but it is extremely versatile. One of the greatest problems facing the gold prospector in Arizona is the lack of water in most of the gold fields. This is also true of the seasonal dredging streams in the Bradshaw Mountains, many of which dwindle to a trickle or go completely dry at certain times of the year. The frustrating part of this is that the gold is still there, but rather unavailable to the prospector. However, he could resort to a shovel and a pan and bring in a few gallons of water to pan out the dirt. That puts him in the same league as a sniper, nothing derogatory intended here, but some prospectors want to mine gold on a bigger scale. Chapter IV tells you how to mine gold using a mini sluice with recyclable water.

There are many of you just starting out in gold prospecting, or, who have been at it for a short time, with only a small amount of money to invest in equipment. As you become proficient in using your gold pan you want to increase your reward. You realize that mining gold with only a pan is just a first step, but would like to process more paydirt with less effort. However, you can't afford the more expensive machines at this time. The answer for you may be a venerable piece of gold recovery equipment that gets gold. This item is so simple, and cheap, that many prospectors build their own. It is a cradle, or rocker. Its history is a long one that goes all the way back to the Fortyniners; and, it is a proven product. Whether you build or buy one, if you use it properly, it won't disappoint you. This is a device that you can take to the dry placers, although you will have to bring along the water for it. Chapter V discusses a cradle and how it may be just the thing for you.

After I published ***Arizona's Golden Secret*** several readers called me to ask if I made and sold drywashers. I was sorry to say that I did not. For those of you handy with tools and not intimidated by working with wood, I have included plans for building a manually operated drywasher. The design is proven and the machine will get gold. Although operated by a pull cord, the principles of gold recovery remain the same for this machine as for the one described in my first book. Just heed my advice about feeding the material slowly, don't overload the riffles, and empty the riffle tray after you process about four buckets of paydirt. If you tire of pulling a cord to dry wash your gravel, you can always buy a crank-operated or motor-powered machine from your local dealer in prospecting equipment. In a dry state such as Arizona, a drywasher is an indispensable piece of equipment and will pay for itself if used often and properly. Chapter VI gives a short course on operating this drywasher.

Classification is the key to efficient gold recovery. In Chapter VII I discuss the need for classification and include plans for building a proven, portable classifying screen.

CHAPTER I

MORE PLACES FOR YOU

The novice prospector always has had a problem finding a place to go to get gold and how to get it. This seems particularly true for Arizona with its small and widely separated gold fields. Many prospectors are reluctant to divulge this information to strangers. From my own personal experience, I realize how difficult it is for newcomers to gold prospecting to discover these closely guarded secrets. One can understand the reasoning behind those already getting the gold to keep out others. Some have the attitude that it's *their* gold and no one else should have it. They keep quiet about it hoping that the newcomer gets discouraged and quits. Others are more open and want to help. My own attitude is that if a person wants to learn something new about things that interest him, he should have that information.

Much of the general information about locations is already published; it is not privileged information. Getting this information may be a little difficult, but, it's there. Many of the gold placers in Arizona are on public domain land, but the *exact* locations of some are a little hard to come by. However, if not already under claim, on patented land, or withdrawn from mineral location, you have the right to prospect for and get gold from these locations. The Federal Government holds these lands in *trust* for the American people. Public domain means just that, for the public, you and me. Congress enacted the mining law of 1872 that specifically gave the American people the right to prospect for, and mine, minerals on public domain land. Although there are many groups and individuals that would like to take these rights away from you and me, they are ours and we must defend them.

I believe that prospecting for gold is a wonderful hobby. We are lucky in that we are in a state that has not only gold, but extraordinary beauty as well. I can't think of a better way to experience Arizona's wonders than on a gold prospecting trip. Just getting to the gold fields is an adventure in itself; and then, there is the keen anticipation that comes as you unload your equipment in preparation for the hunt to follow. In a way, prospecting for gold is like a hunt for big game. With gold, we are dealing with a quarry that is elusive and difficult to bag. We are pitting our skill and knowledge against a formidable adversary, Mother Nature, herself. For some, the thrill of the hunt is the reward, for others, the kill. Whatever is the reward, each of us that goes for the gold experiences the raw beauty of nature in a way that few people ever do.

The gold placer locations I described in ***Arizona's Golden Secret*** are places that anyone can get to, and find gold. I personally tested and mined these locations and got gold from them. Some places have bigger gold nuggets than others but that doesn't mean that they are better than others. To get the big nuggets requires considerably more effort, and *luck*, on your part. Remember: Whether large or small, gold is **gold!**

As the more accessible and popular gold placers become crowded with prospectors, some of you would like to try other areas. Many of these are little-known places and off the beaten path. Some are close to population centers, others are very remote; but, all have gold. For these adventurous individuals, I include this chapter. In it I describe locations that few prospectors visit. Some placers have seen very little mining since the depression years of the 1930s. Several are difficult to get to. These are challenging placer fields. All require skillful application of your knowledge of finding, and mining, gold. I can't promise you that your monetary reward will be great; but, I do think that your spiritual one will be.

I have tried to give you as much detailed information as I can for you to find these fields and what to expect when you get to each one. This is the best I can do. The rest is up to you.

NORTH OF TUCSON PLACER LOCALITIES (Fig. 1):

There are two gold districts, often lumped together, north of Tucson near the town of **Oracle.** The **Cañada Del Oro District** lies southwest of Oracle (Figs. 2, 2A, 2B, 3 & 4). Spaniards reportedly found gold in Cañada Del Oro as early as 1698. For the next 150 years Spaniards and Mexicans worked the gravel under constant threat of attack by the Apache. Americans arrived around the middle 1850s, and they also worked the gravel. They, too, were subject to attack by the Apache until the official quelling of the Indians in 1885.

The **Old Hat District** lies southeast of Oracle (Figs. 9 & 10). Prospectors found lode gold deposits in the late 1870s. In 1878, Isaac Lorraine discovered a lead-silver lode deposit that he called the American Flag, and in 1881 the wife of prospector Capt. John T. Young discovered the Southern Belle lode gold deposit. Prospectors found placer gold in Southern Belle Wash below the Southern Belle Mine shortly thereafter. Mining for lode gold and placer gold continues in both districts to this day.

Those wanting to mine placer gold in Cañada Del Oro's upper reaches will find access difficult. A look at the map shows a road heading southeast from State Highway 77 to the upper reaches of Cañada Del Oro wash but this is a bit misleading. It branches off the main paved road that leads to Biosphere 2, about 1-3/4 miles south of the State Highway, but active lode mining operations block all entry. Signs proclaim the area east of the Biosphere 2 road as private property and warn against trespass.

The lower, southern portion of Cañada Del Oro wash is more accessible, but much of it is in private ownership. A small portion is state-owned. To reach it, drive north on Oracle Road to Golder Ranch Road, near the town of **Catalina.** Go east on Golder Ranch Road to Lago Del Oro Parkway, then north on Lago Del Oro Parkway to Edwin Road. Head east on Edwin Road to Santa Monica Trail. There is a wooden sign at the end of Edwin Road reading "Charouleau Gap," and an arrow on it pointing north. Take the dirt road to your left, which is Santa Monica Trail, to the wash. If you cross the wash you may continue on the dirt road signed "Hielo" then take the second left-hand dirt road marked "736." This is the Charouleau Gap road. It will take you to the headwaters of Cañada Del Oro wash but you must have a four-wheel-drive vehicle to negotiate it. It is very rough. If you head north on the first left-hand dirt road, Santa Monica Trail, an active sand-and-gravel operation blocks the way about a half mile upstream. Watch out for deep sand threatening to entrap your vehicle along this route!

The larger gold in Cañada Del Oro wash lies under tons of cobblestones and boulders brought down from the mountains by recurring flash floods. Bedrock is *very* deep. I confined my mining operations to high banks (Figs. 5, 6 & 7), gravel bars and lower benches. Although the wash flows for most of the early spring months this is primarily a "desert placer." Gold occurrences are spotty, and paystreaks are erratic. The deposition of gold under flash flood conditions accounts for this. Mining for gold in the Cañada Del Oro district, at this lower elevation, is a hit-or-miss proposition.

The gold I found was very small, ragged grains, ranging up to one-twentieth of an inch in diameter. However, not all tests produced gold. A test on a bedrock exposure was barren (Fig. 7). However, high banks and benches cut by flash floods of more recent times reveal layers deposited long ago. It is possible that other layers (indicated by coarse gravel streaks) may carry better gold. Black sand is very abundant in samples taken from these gravel streaks, which may also contain gem-quality garnet. Although I high-banked about one cubic yard of bench gravel in late April 1993, my returns were modest due to the erratic paystreak, and the relatively inefficient sluice design I used for this size of gold. To get any appreciable amount of the very finely divided gold, you have to process *many* cubic yards of material, and have a recovery system capable of extracting it. The small size of the gold mandates that you save all black sand concentrate for further processing.

In spite of the small size of the gold, lower Cañada Del Oro is a very short drive from Tucson, easy to get to and easy to mine. Presumably, coarser gold occurs on the headwaters but finding it will be just as challenging. Cobblestones and boulders hinder your efforts; and, the paystreaks are sporadic and spotty. However, the natural drainage area of Cañada Del Oro is very extensive, and opportunities for panning gravel from many different situations abound throughout its length. Due to the presence of abundant black sand, and very fine gold, I consider the lower portion of the district an excellent one to perfect your gold-panning techniques. Summertime temperatures are very high here; your best times are the fall, and early spring, months.

Getting to the Old Hat District is much easier (Figs. 9 & 10). Drive north out of Tucson on Oracle Road to State Highway 77, then east on 77 to the business route turnoff to the town of Oracle. Go east on the business route and take the **Mount Lemmon** road eastward from the intersection; do not take the Mammoth-Winkleman road through Oracle.

Continue on the Mount Lemmon road beyond the mileage sign indicating 5 miles to **Peppersauce Campground** and take the graded dirt road heading south toward the mountains. This is the Mount Lemmon road that eventually goes up the north side of the Santa Catalina Mountains to near the summit. The stretch of road south to Nugget Canyon traverses the Old Hat mining district. There are many parcels of private and patented lands as well as valid lode and placer claims in the district. Placer miners have active, valid claims on all of Southern Belle Wash, the best wash for placer gold. My observation of one placer mining operation revealed the gold in this wash as coarse and many flakes exceeded one-tenth inch in diameter. The placer miners did not reach bedrock in that locality.

My introduction to **Peppersauce Wash** came in 1967 and I have visited this locality several times since. I conducted most of my placer testing in the wash upstream from the campground. The wash has a reliable flow very early in the year, allowing you to use a small suction dredge, but becomes dry by mid-April in the lower reaches. A small, dependable flow, from springs, continues indefinitely in the upper portion. Throughout its length, the wash contains a hodgepodge of cobblestones and boulders that make mining difficult and arduous.

In March 1993 I operated a 3-inch suction dredge on the lower portion of the wash, about one quarter mile west of the campground (Fig 8). I found much fine gold, and several coarse flakes, in a streak approximately two feet below the wash surface, and in the side bank. Later, in April, I returned twice to the same spot in the company of my wife, Alyn, and two good friends, Mike London and Debra Torres of Tucson, Arizona. Mike and Debra operate a gold prospecting store and are enthusiastic prospectors.

I made several test excavations, using my portable grizzly screen to classify the gravel. Excavation was laborious and slow due to the many cobblestones and boulders. For excavating the gravel, I used a pick-mattock, hexagonal steel digging bar and a sharp-pointed shovel. I did most of the digging with the digging bar. Mike and Debra processed the classified gravel through a mini sluice using self-contained, recycled water (See Chapter IV).

My tests confirmed that Peppersauce Wash contains fine, and coarse, gold in discontinuous, irregular streaks, not unlike Nugget Canyon farther south. Most gold occurred not more than one to two feet below the surface of the wash and going deeper was not necessarily better. I came across a clay layer at about three feet below the surface of the wash. It was barren. Several hundred feet upstream I tested gravel behind some very large boulders in the middle of the wash. The gold occurred in much the same situation, about one to two feet below the surface of the wash. One small nugget, several flakes and much fine gold came out of this layer. Below it, I came across the barren clay layer. Overall, the gold is bright, rough, and flat, with much oxide (indicating a close source), and is generally small. Few particles exceeded one-tenth of an inch in diameter. However, your chances are good for finding larger ones almost anywhere in the wash as a result of violent stream bed movement caused by flash floods.

Peppersauce Wash is on the southern fringe of the Old Hat District. Its flow is intermittent and subject to flash floods from torrential rains. The tremendous amount of intermixed cobblestones and huge boulders, and the erratic paystreaks in its upper layers, indicates that this wash is a "desert placer." Typical stream placer conditions do not apply. Gold occurs in a spotty fashion as a result of deposition during flash floods and is a hit-or-miss proposition, much like Cañada Del Oro. Generally, you will have more success by prospecting the layer from one to two feet below the surface of the wash. Try to stay out of the clay layer itself, which I found lacking in gold. The clay layer, presumably extremely fine silt (flour) deposited at the end of a flood period when flow slackens, acts somewhat as a false bedrock. I did not attempt to penetrate it since it meant digging a hole exceeding three feet in depth; I felt that the character of this placer did not justify the effort.

In mid-May 1993 my son, Roger, and I sniped bedrock exposures upstream from my previous tests and obtained good color from cracks and pockets. Sniping the bedrock exposures is one of the best ways to mine gold in Peppersauce Wash since it involves moving only a minimum of overburden. Try to clean out the cracks, most of which have trapped gold washing downstream during flood conditions.

Peppersauce Wash is very easy to get to in a passenger car. The U. S. Forest Service maintains a fine campground lower down with drinking water and pit toilets. This locality in the Coronado National Forest is over 4,000 feet in elevation and heavily wooded. If you can, avoid going there during the rainy months of December, January, July and August. It is a favorite place for bird watchers and naturalists. Those who come to mine its gold do so in beautiful surroundings. I found this experience, in the company of my wife and good friends, as rewarding as the gold itself.

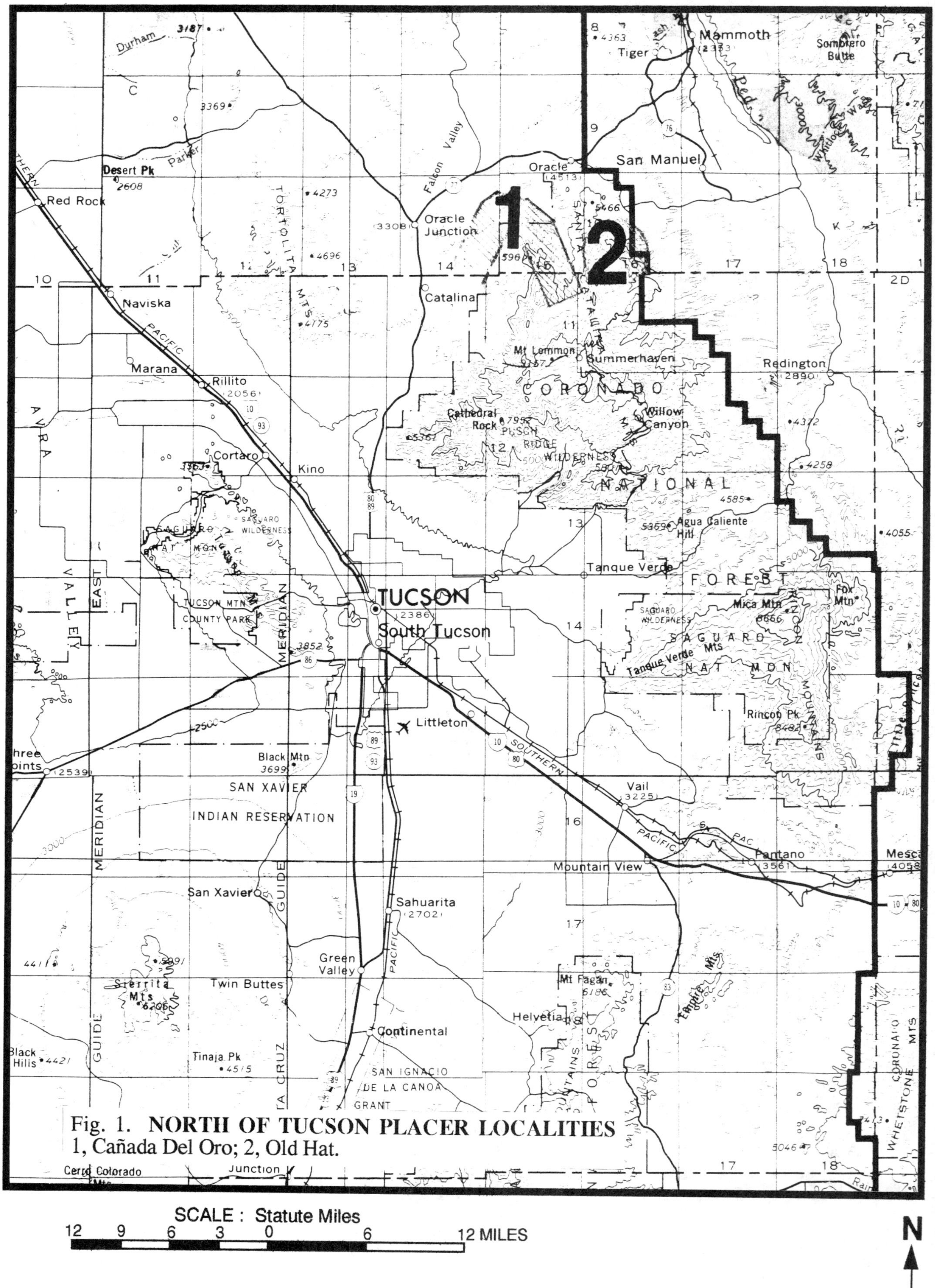

Fig. 1. **NORTH OF TUCSON PLACER LOCALITIES**
1, Cañada Del Oro; 2, Old Hat.

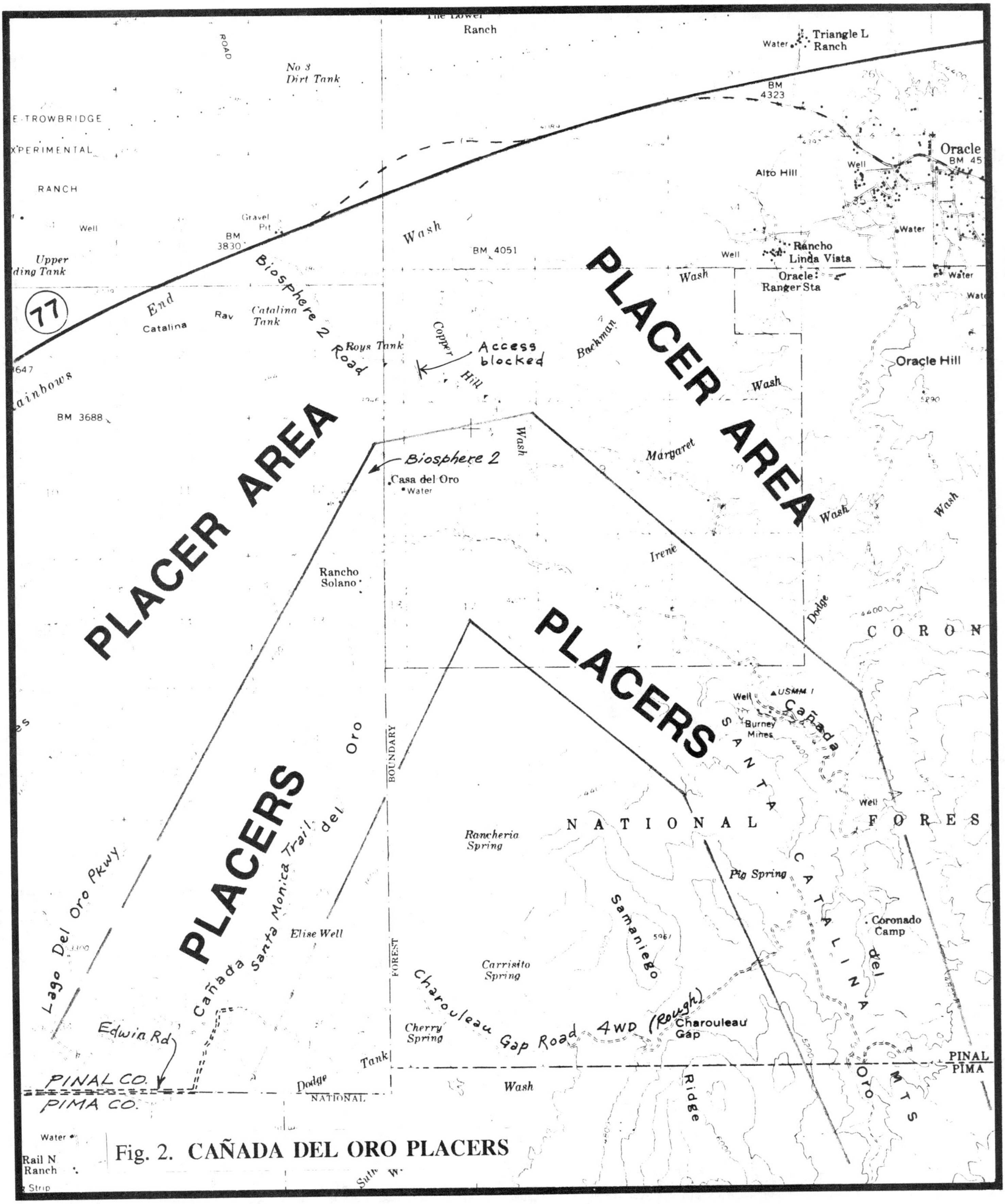

Fig. 2. CAÑADA DEL ORO PLACERS

SCALE 1:62500

1 1/2 0 1 2 MILES

N

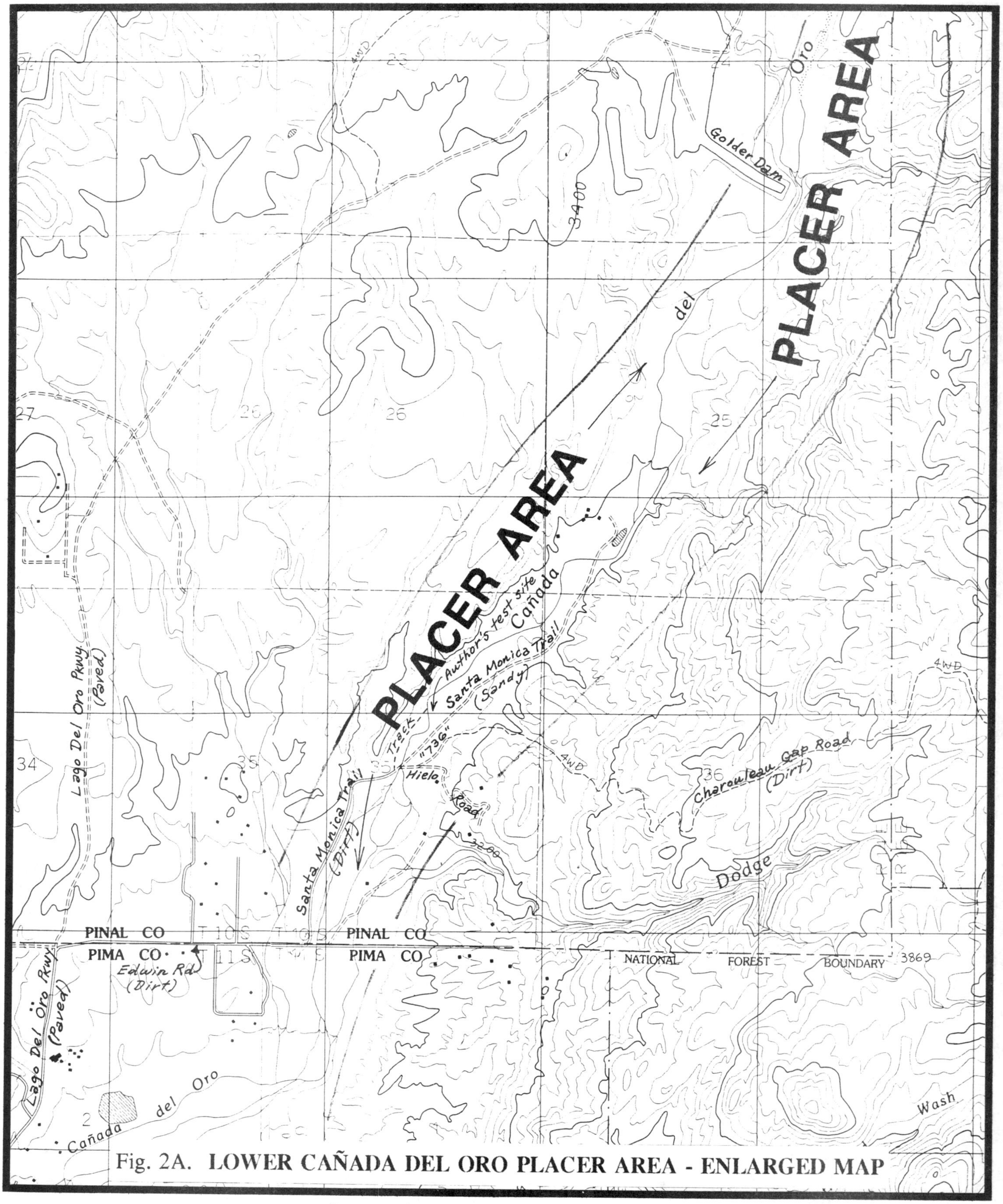

Fig. 2A. **LOWER CAÑADA DEL ORO PLACER AREA - ENLARGED MAP**

SCALE 1: 24000

1 — 1/2 — 0 — 1 MILE

N ↑

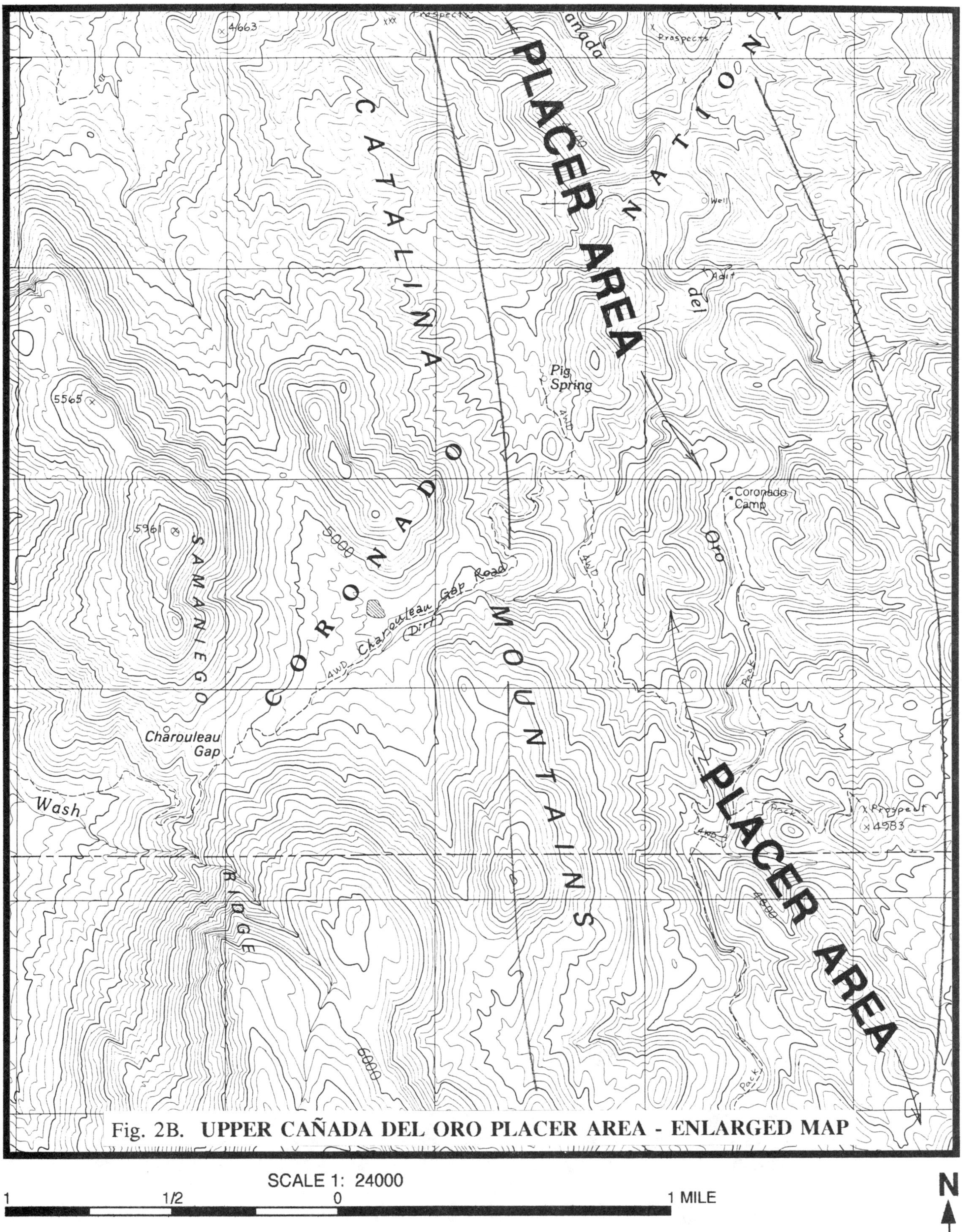

Fig. 2B. UPPER CAÑADA DEL ORO PLACER AREA - ENLARGED MAP

Fig. 3. **CAÑADA DEL ORO WASH NEAR CATALINA, ARIZONA, LOOKING TO THE NORTHEAST**. Arrow indicates bench gravel high-banked, and also drywashed, by the author, April and May 1993. Fine gold occurs here in erratic, discontinuous streaks associated with coarser gravel layers.

Fig. 4. **CAÑADA DEL ORO WASH NEAR CATALINA, ARIZONA, LOOKING TO THE SOUTHEAST**. Arrow indicates very high bank at the end of the dirt road in the foreground tested by the author, April 1993 (see Figs. 5, 6 & 7).

Fig. 5. Author excavating a high bank gravel streak for testing, Cañada Del Oro District near Catalina, Arizona, April 1993. Very fine gold and abundant black sand occurred in this layer. Note irregular layering of gravel within the bank indicating flood deposition.

Fig. 6. Author pointing to two gravel layers just above exposed granite bedrock upon which he crouches, Cañada Del Oro District near Catalina, Arizona, April 1993.

Fig. 7. Author excavating gravel layer resting on granite bedrock for testing, Cañada Del Oro District near Catalina, Arizona, April 1993. This layer was barren of gold.

Fig. 8. Author suction-dredging Peppersauce Wash, Old Hat District, March 1993. Note low stream flow, and abundant cobblestones and boulders characteristic of a desert placer subject to flash floods. Fine gold and coarse flakes occur in a streak approximately one to two feet below the surface of the wash at this locality.

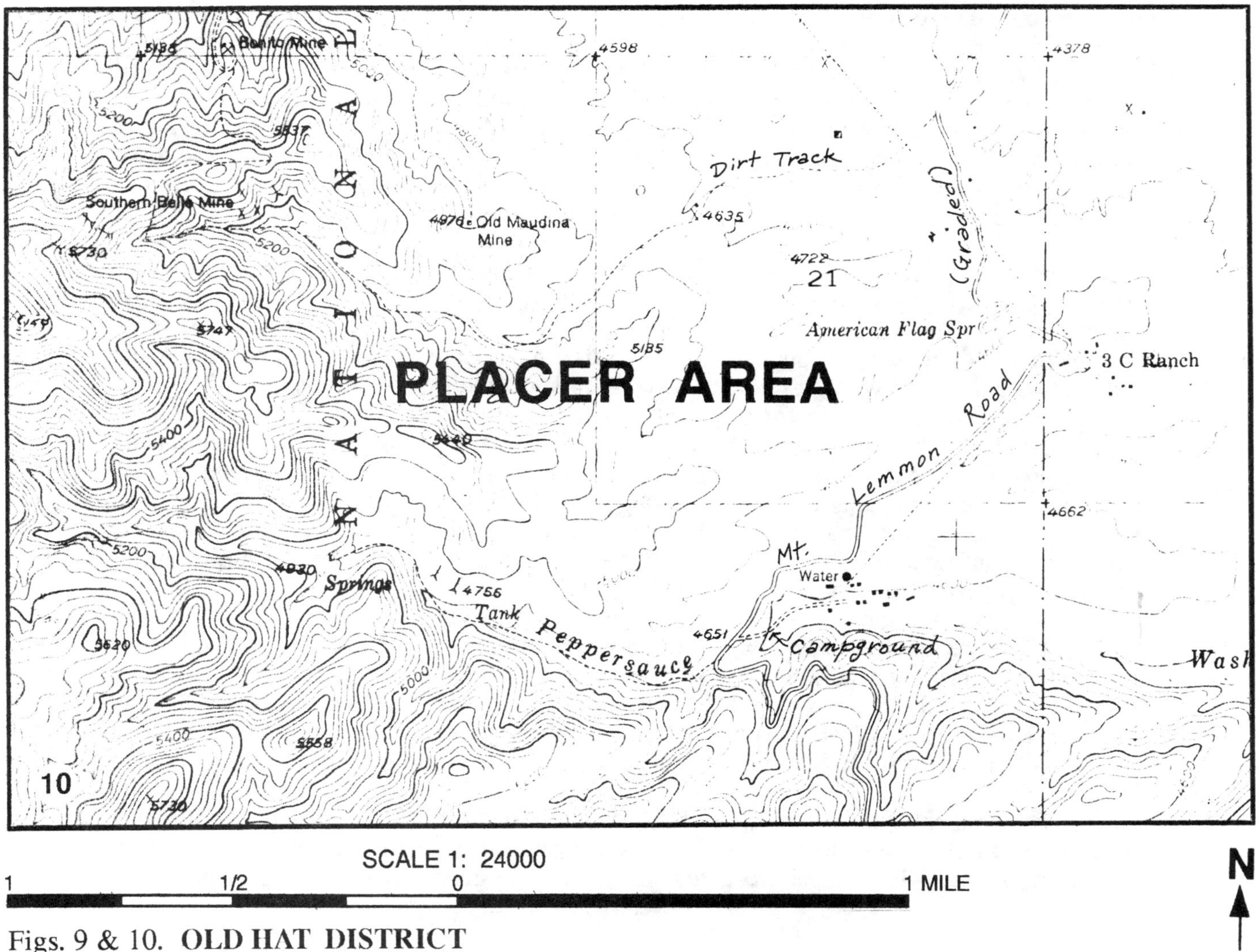

Figs. 9 & 10. **OLD HAT DISTRICT**
9, Dirt track heading west to Southern Belle Wash; 10, Enlarged map of major placer area.

SOUTH OF TUCSON PLACER LOCALITIES (Fig. 11):

In my book *Arizona's Golden Secret*, I reported on the Las Guijas, Oro Blanco, San Luis Mountains and Greaterville placers. The next three placer fields round out my coverage of southern Arizona's placer districts. There are several-more remaining not available to the recreational prospector in that they occur on private, or Indian reservation, lands.

The **Patagonia Mountains** lie to the east and northeast of **Nogales**, Arizona (Figs. 12, 13, 14 & 15). This small, rugged range of mountains in Santa Cruz County reaches a maximum height of 7,221 feet on Mt. Washington, and receives much precipitation in the form of rain and snow in the higher parts. The lower elevations have a heavy growth of oak trees and a shrub under-story; the higher elevations have juniper and pine forests. In wet years some of the washes flow for a few months, but most remain dry throughout the year. During the summer rainy season chiggers infest the lush vegetation! If you are susceptible to these nasty creatures, use **liberal** applications of insect repellent from head to toe beforehand. No part of your body is safe should you venture into the brush without this protection!

Although the Spaniards entered the region as early as 1539, we have little evidence that they mined here. Our earliest evidence comes from the remains of mining by Mexicans shortly after 1821. Americans arrived in the district around 1854. Lt. Sylvester Mowry of the U. S. Army bought the "Patagonia Mine" in 1859, which later became the Mowry Mine. Prospectors located other lode mines, and the mining camps of Duquesne, Harshaw, Mowry and Washington Camp sprang up. A few rugged individuals make Washington Camp home today, but most camps are deserted ghosts. The historic old town of **Patagonia** merits a visit by those wanting to savor the flavor of a slower-paced, authentic Western way of life so attractive to the many retirees settling there.

Most of region is in the Coronado National Forest. However, there are many patented claims and private parcels within the boundaries, particularly on the east side of the mountains. The western side of the mountains, facing Sonoita Creek, is more open for prospecting and virtually untouched. Some roads are passable in a standard passenger car but most require four-wheel-drive due to washouts, or muddy conditions, resulting from heavy rains. The best times for prospecting are the dry periods from October through November and again from late February through June. During these periods the region has a very pleasant climate. However, the time of the summer rains, usually the first part of July through September, is very uncomfortable. With the arrival of the rains, higher daytime temperatures, and extremely high humidity, this region takes on a tropical aspect. Much of the mountains' flora and fauna reflect this tropical influence so unique to the southeastern part of Arizona.

There are inactive lode mines throughout the mountains. Most of the lode mines in the district were silver-lead-zinc producers, except for the Three-R copper mine. Virtually every wash in the district carries gold. Most of the gold is small and associated with abundant black sand. Sizable nuggets are very rare, although you could get lucky. The vast, western side of the district, which saw relatively little mining, is your best bet.

The **Nogales**, or **Gold Hill** (Mt. Benedict), district is little-known. Although it is only three miles from the city of **Nogales**, Arizona, it is small and relatively obscure (Figs. 16, 17, & 18). There are a few inactive lode gold mines on and around Mt.

Benedict. Spaniards discovered rich gold ore in the district shortly after the establishment of the Guevavi Mission, close to the Santa Cruz River, in 1691. They worked placer gravel nearby, probably Guevavi Wash, and undoubtedly worked some of the washes draining the eastern and northeastern flanks of Mount Benedict. Americans took over the region in 1853. The period from 1853 to the late 1870's was a dangerous one for prospectors who were under constant threat of attack by the Apache. After the cessation of hostilities, mining activity peaked around the turn of the century and continued, sporadically, until about 1967.

Most of the area is a mix of stock-raising homestead lands, with the surface in private ownership and the Federal government administering the mineral rights. There are state-owned lands in parts of the district, and some privately owned parcels not open for prospecting. The Santa Fe Ranch posted no-trespassing signs all along the eastern boundary fence advising against intrusion upon its land, at any time, for whatever purpose. Additional signs also proclaim this area a "wildlife refuge." I entered another sector of the placer area through an unlocked gate at the northwestern boundary. There are no warning signs anywhere at this entry or in the immediate vicinity. You may want to check land status records at the Santa Cruz County Courthouse in Nogales before you do any mining here.

To reach the district drive south from Tucson on I-19 to the Ruby Road Interchange. Exit I-19 at the interchange and go east to the frontage road. Take the frontage road south 0.4 mile to Old Tucson Road. Go east, then south, 0.6 mile on Old Tucson Road to South River Road, and take South River Road eastward 1.1 miles to the unlocked (at the time of my visit) gate to your right. South River Road is a paved road to the gate entry. The road southward through the entry is dirt. This is the road to the water tank, and the Gold Hill Mine. You can negotiate it in a passenger car using care in a few places. It will take you beyond the water tank to a locked gate that prevents you from driving farther. There are no signs on the gate or the fence line prohibiting entry. The Gold Hill Mine is a short hike from here (Fig. 19).

I test-mined a small wash crossed by the dirt road (Fig. 20). It is south of the first major wash. Test pans revealed some colors of gold associated with black sand streaks. This encouraged me enough to process several buckets of bedrock gravel through my drywasher. Black sand is so abundant that after four buckets the riffles clogged. The concentrate carried very fine color no bigger than the grains of black sand. Selected test-mining in prime spots proved a little disappointing. In some cases, gold was absent. A small wash such as this one is subject to flash-flooding, and scouring, from torrential rains. This may account for the small gold size and erratic paystreaks. Bedrock occurs less than a foot below the surface virtually throughout this wash, and others in this district, and crops out in many places (Fig. 20). For the most part, bedrock is rough and its outcrops fissured. The wash itself contained only a few scattered cobblestones and boulders. I consider this excellent terrain for drywashing, sniping and crevicing. Concentrate from my drywasher contained a number of buckshot, and I noted the presence of empty bullet cartridges here and there. If you plan on metal-detecting the area, these are the items you have to contend with; I did not come across any nails or other small ferrous objects.

Washes available for panning in this district are small and few. To get to their heads, particularly on Mount Benedict itself, requires hiking. A thick growth of catclaw mimosa and mesquite hinder your efforts at cross-country hiking (Fig. 19). Bedrock exposures abound in this highly mineralized district. I believe that this field merits a closer look, particularly in washes draining away from the lode mines. The fine gold I found came from considerably lower down and larger, coarser flakes and nuggets probably occur in the upper reaches. In spite of this district's relatively small area and proximity to Nogales, I did not see, or meet, anyone else.

The **Sierrita Mountains** in Pima County are approximately thirty miles south-west of **Tucson** (Figs. 21, 22 & 23). This is a low, desert range rising to 6,188 feet on the top of Keystone Peak, which is the highest point. Pediment surfaces are very broad, gently sloping, and extend for miles beyond the mountains.

Spaniards discovered lode gold and silver deposits as early as 1720. Americans entered the area in the middle 1850s. Until the stopping of Apache Indian attacks in the late 1870s, there was very little mining done in the district. In 1955 the Pima Mine, on the eastern side of the mountains, opened as an open-pit copper mine. Other, large open-pit copper mines soon followed.

This highly mineralized range is a patchwork of ownerships that make access somewhat difficult. The large copper mines and the Sierrita Mining and Ranching Company control vast portions of the mineralized areas. Access to the higher reaches of the eastern side is through a single locked gate at the McGee Ranch Settlement. The western side is more accessible although there are many private parcels mixed in among the state trust lands that predominate. Only a small portion of the western part of the range is public domain land, mainly around Stevens Mountain. There are many lode mines and prospects throughout the Sierritas. Dirt tracks and ranch roads crisscross the western side; most, if not all, require vehicles with high centers or four-wheel-drive, particularly near the mountain flanks. Locked gates block access to many roads. Before attempting entry into this part of the district, you should obtain the latest USGS 7.5' quadrangles. These maps show almost all the dirt tracks and roads, but do omit a few. Without the latest maps to guide you, getting lost here is a real possibility. You can enter the district on three dirt roads heading southeastward from State Highway 286. They are (1) the road to Stevens Mountain paralleling Fresnal Wash; (2) the road along Banner Wash to the Banner Mine; and (3) the road along Ash Wash to Tascuela Wash and San Juan Wash.

To get to Stevens Mountain, take the first left-hand turnoff (if you are heading south on Highway 286) just a few hundred feet south of the Anvil Ranch Road. The gate does not have a lock but remember to close it behind you. There are other, unlocked, gates beyond. This road, as well as others in the district, is rough and goes in and out of deep, sandy washes. The Banner Mine Road is the Wild Horse Ranch Road, 0.7 mile south of Mile Post 35. You can reach the Banner Mine area from the Stevens Mountain Road and avoid going through the Wild Horse Ranch. The road along Ash Creek to the Sunshine Mine is just north of the Palo Alto Ranch. The private land owner has granted access at will; meaning, he can take it away at any time. In the early days, Ash Wash just below the Sunshine Mine was a good producer of placer gold.

I have drywashed the area around Stevens Mountain and found much fine gold that carries a coppery tint. In many of the washes the gravel is caliche-cemented and loose gravel is rather thin. The larger washes have huge volumes of gravel but the gold is spotty and streaky. I consider all washes in the district the "desert placer" type making for erratic paystreaks and difficult prospecting. These washes remain dry most of the year but are subject to flash-flooding from heavy rains. Summer temperatures here are very high.

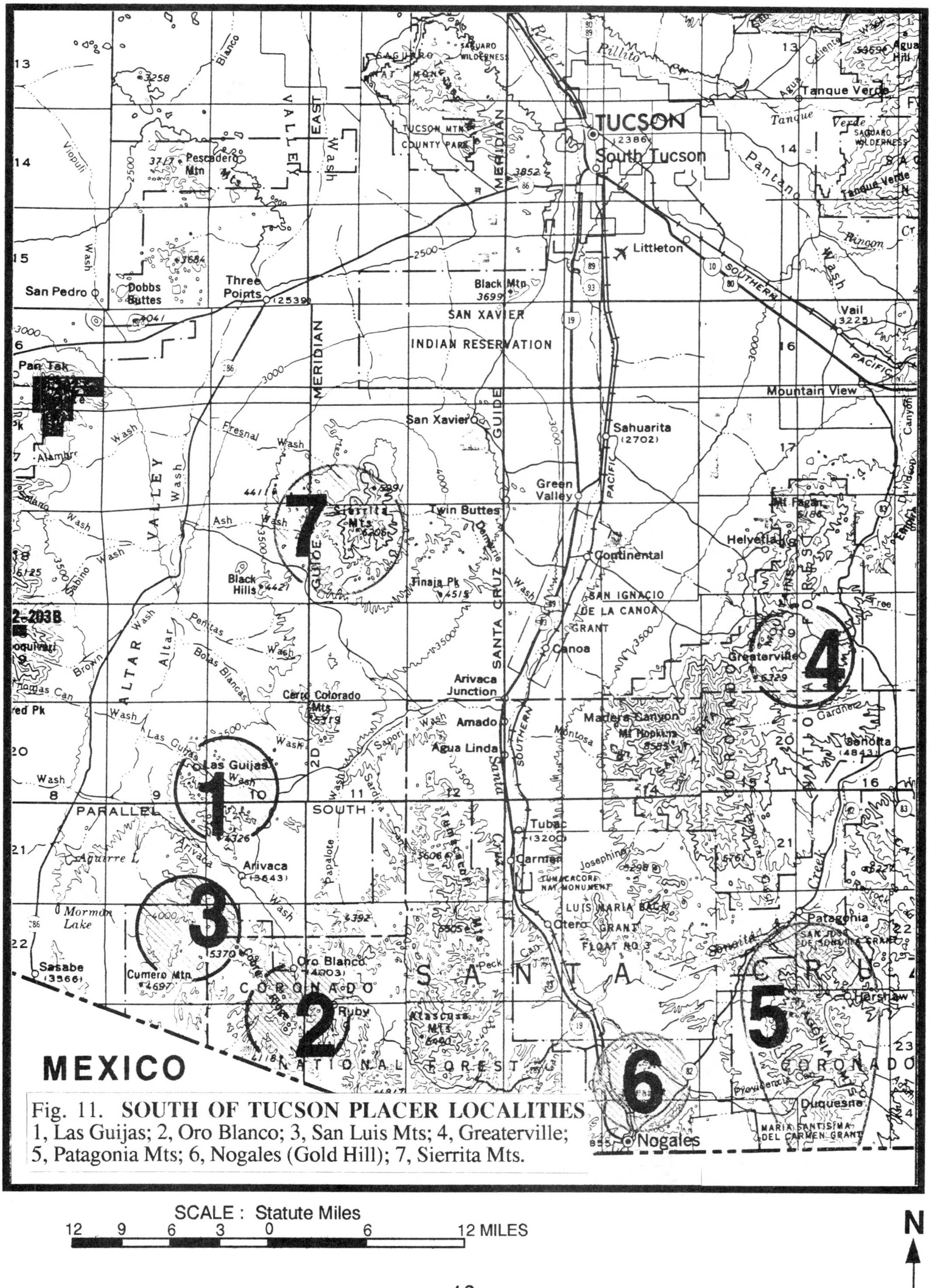

Fig. 11. **SOUTH OF TUCSON PLACER LOCALITIES**
1, Las Guijas; 2, Oro Blanco; 3, San Luis Mts; 4, Greaterville;
5, Patagonia Mts; 6, Nogales (Gold Hill); 7, Sierrita Mts.

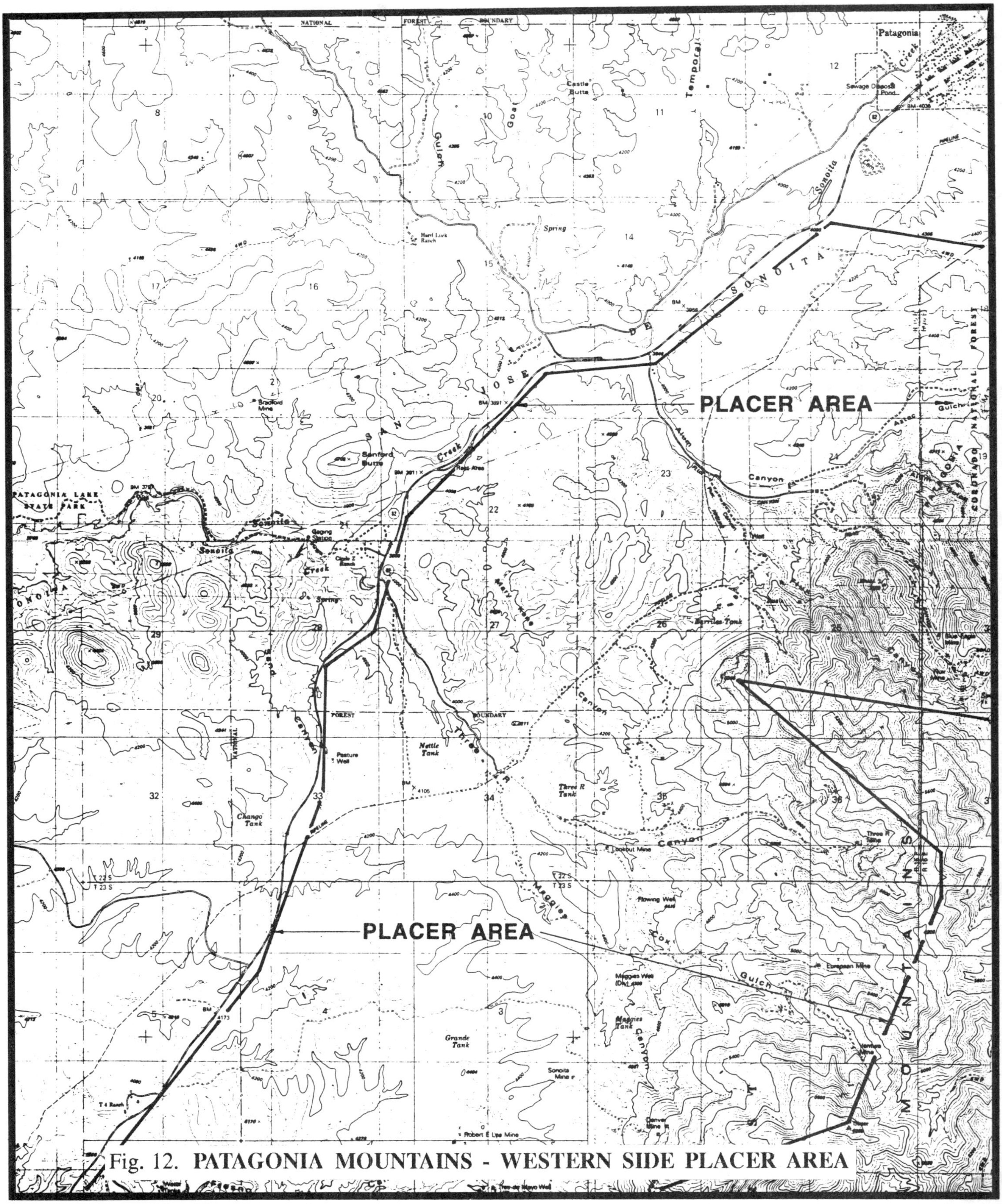

Fig. 12. PATAGONIA MOUNTAINS - WESTERN SIDE PLACER AREA

SCALE: Statute Miles

1 1/2 0 1 2 MILES

N

Fig. 13. PATAGONIA MOUNTAINS - EASTERN SIDE PLACER AREA

SCALE: Statute Miles

1 1/2 0 1 2 MILES

N

Fig. 14. Patagonia Mountains - western side - as seen from State Highway 82, June 1993. Virtually untouched placer ground exists on this side of the mountains, but access is difficult due to private drives and no-trespassing signs. Your best bests are Flux Canyon Road (#812), Three R (#215) to Maggies Canyon (#235, which loops back to Highway 82), and Wild Hog Canyon (#125) from Duquesne Rd.; all are 4WD.

Fig. 15. Author panning for gold on Harshaw Creek, Patagonia Mountains, June 1993. This creek almost always has water in it along certain favorable stretches sufficient for the operation of a backpacker type of suction dredge. There are many bedrock exposures. The gold is small and accompanied by abundant black sand.

Fig. 15a. Author's four-wheel-drive truck parked in the shade of a live oak tree, looking down Guajolote Wash, Patagonia Mountains, June 1993. There are many uncrowded spots available for camping such as this one throughout the Patagonias.

Fig. 15b. Author excavating gravel from a small tributary of Guajolote Wash, eastern side of the Patagonia Mountains, June 1993. The gold I found here is a light yellow, ranging up to 1/20 inch in diameter. It occurs throughout the gravel, but concentrates mostly on the cemented false bedrock.

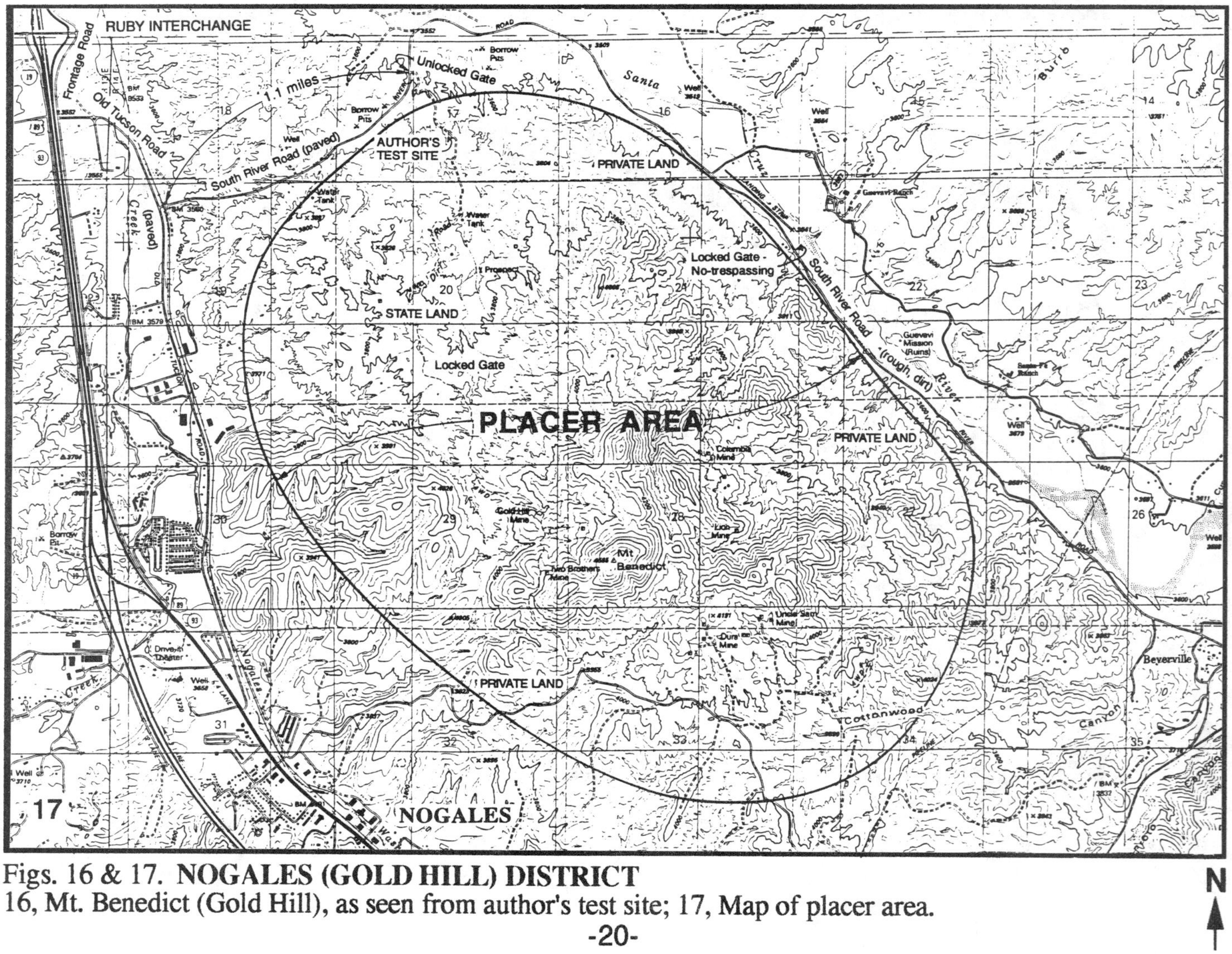

Figs. 16 & 17. **NOGALES (GOLD HILL) DISTRICT**
16, Mt. Benedict (Gold Hill), as seen from author's test site; 17, Map of placer area.

N

Fig. 18. **NOGALES (GOLD HILL) DISTRICT - ENLARGED MAP**

Fig. 19. Mt. Benedict (Gold Hill), Nogales District, as seen from the Gold Hill Mine road past the locked gate, looking to the Southeast. Predominant vegetation is mesquite trees and catclaw mimosa shrubs.

Fig. 20. Author testing a small wash with his drywasher, Nogales (Gold Hill) District (see Figs. 17 & 18). Note bedrock exposures on both side banks in the foreground (arrows). Overburden here varies from only a few inches to over one foot and contains few cobblestones and no boulders.

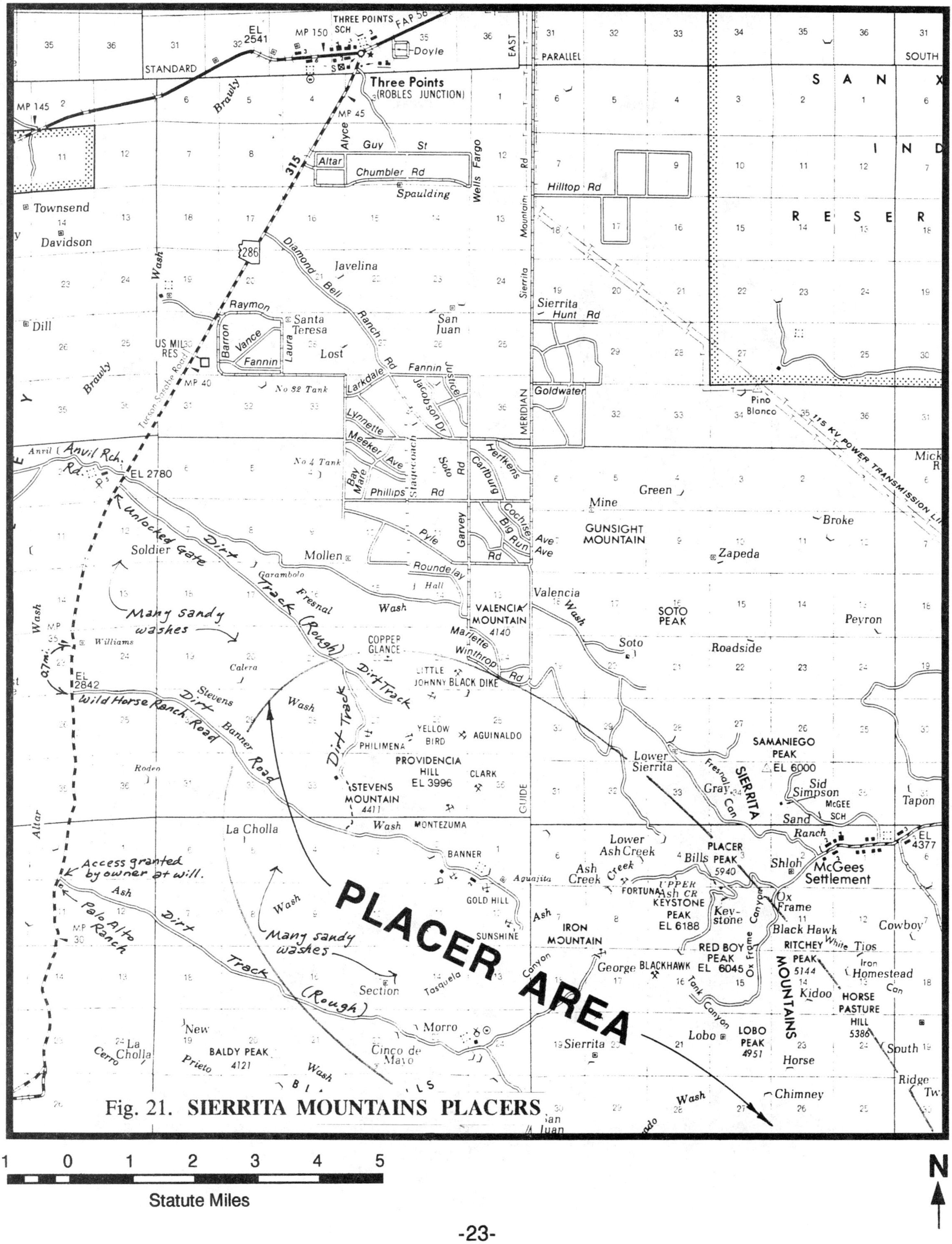

Fig. 21. SIERRITA MOUNTAINS PLACERS

Fig. 22. Sierrita Mountains, as seen from Diamond Bell Ranch Estates, looking to the Southeast.

Fig. 23. Sierrita Mountains, western side. Typical dirt road heading to the mountains. Access roads to the placer areas near the mountains require high-centered, or four-wheel-drive, vehicles. Very few dirt roads in this district receive maintenance.

BRADSHAW MOUNTAINS, PRESCOTT AND VICINITY (Fig. 24):

I covered Lynx Creek, Hassayampa Creek, Big Bug Creek, Turkey Creek, Weaver Creek, Vulture Mountains, San Domingo Wash and Soap Creek in ***Arizona's Golden Secret***. All these places have good gold and are easy to get to. The next two placers are a little trickier to find.

The **Humbug Placers** center on the ghost towns of **Humbug** and **Columbia** situated on Humbug Creek (Fig. 25). Much of the gold found in this district came from Rockwall, Swilling and Carpenter gulches, tributaries of Humbug Creek. In 1980, my brother Joseph and I prospected the small washes in the area. All carried abundant black sand and bright fine gold. Bedrock outcrops are common enough to afford many opportunities for sniping and crevicing. In spite of the district's remoteness, there are reclusive persons living in the area as well as ranchers. These private landowners have posted their properties clearly against trespass.

In recent years homicides have become common in the area. Due to its remoteness and extremely rugged terrain, it is a dumping ground for victims of murders occurring elsewhere, and for opportunistic murder and robbery of unprotected persons just being alone here. A Yavapai County worker, that I happened to meet here recently, carried a sidearm for his protection. As the popular credit card commercial goes, he said that he "Doesn't leave home without it." If you decide to enter this area to prospect, **heed his advice**! In remote areas of Arizona such as this one a firearm is your best protection.

When my brother and I visited the area we were able to drive directly to Humbug and Columbia by way of the dirt road through Castle Hot Springs. Then we drove through Champie Ranch, up along Cow Creek, and east to the ghost towns. Now, a locked gate blocks this way at the VX Ranch north of Champie. To get to the district, today, you must take the **Crown King Road** north of **Lake Pleasant**. From I-17, going either north or south, take the paved Lake Pleasant Road westward. Turn north on Castle Hot Springs Road and stay on it. North of the turnoff to the lake the paved road becomes dirt. Continue on it to where it crosses a large wash. Floods from heavy rains may make this wash impassable. After you cross the wash you come upon an intersection with a sign posted on the right-hand road reading "Crown King." Take the Crown King Road.

If, however, you take the left-hand road you will wind up at Castle Hot Springs, and on the way there, drive through and cross Castle Creek many times. You can get stuck here, particularly in times of flood. This was the way we went many years ago. About a mile west of Castle Hot Springs the road forks. The left-hand fork goes to Morristown; the right-hand road goes to the Champie and VX Ranches. Since the locked gate controls access to Humbug and Columbia this way, you must take the Crown King Road.

Go north on the Crown King Road until you come upon three homes on the hill to your left. Two of these are abandoned and the third one is occupied (see Figs. 25a & 25b). A hand-painted sign on the lid of a 55-gallon drum reads "Crown King" with an arrow pointing to the left. On your right there is an obscure dirt road leading down to the wash that skirts the remains of an abandoned homestead. This is the Columbia Pack Trail. It crosses the wash, Cow Creek (Fig. 25c). If you are a very skilled four-wheeler, take the left-hand dirt road at the fork (Fig. 25d) and go on to Columbia. This road is very rough! If you prefer to hike, Columbia is about 3 miles from here. Take plenty of water with you.

Cow Creek also offers opportunities for gold prospecting. Instead of going on to Columbia you may want to try your luck on this creek. At the time of my visit the creek had water running in it. Bedrock exposures abound along the creek and offer many possibilities for sniping. I worked bedrock pockets and found some coarse gold along with abundant black sand and garnets. Although this area is private property, the Federal Government owns the mineral rights. In the Humbug and Columbia locales the land is public domain administered by the BLM. However, the townsite of Humbug is private land. There are patented lode mines in that area as well as valid lode claims, particularly in Section 12. Check out land status records first before you do any mining here.

Remember: This is a very remote, rugged and dangerous terrain to visit, especially so in the summer months as daytime temperatures are very high. I do not recommend going alone. You must prepare yourself for desert survival if there is a vehicle break down. If your vehicle does break down and you leave the area to seek help, don't be away from it longer than 24 hours. The county worker told me that it *will* be stripped after that, perhaps even sooner.

In the late 1970s a small group of suction-dredgers from Phoenix regularly went up to **Copper Basin** near **Prescott** to mine gold on Phelps Dodge property (see Figs. 26 & 27). Then, they could do so. However, times have changed. You may not mine gold on Phelps Dodge property now, <u>or on any other private land in this area</u>. Well-posted signs mark private property here. I conducted my mining on Forest Service land, away from the private holdings and off any valid claims. Claims status here changes from year to year.

There are two easy ways to get to Copper Basin from Phoenix. You can go North on I-17 to **Cordes Junction**, then take State Highway 69 to Prescott, which merits a visit or a stay. From Prescott, take **Copper Basin Road** and follow it to Copper Basin, a distance of about eight miles. This is a good graded road passable in a passenger car.

You also may get to Copper Basin by way of **Wickenburg**, **Yarnell** and **Wilhoit**. From U. S. Highway 89 north of Wilhoit, turn west on Forest Road **53**. A numbered sign marks this road that is rough in spots, particularly where it comes up to meet the Copper Basin Road. The creek on your left is Copper Creek. This creek drains eastward and is not the same as Copper Basin Wash. It has water in it for most of its course and carries small, coarse gold flakes and abundant black sand. There is enough water in the creek to operate a small backpacker type of dredge as well as a mini sluice. There are many opportunities along Copper Creek for the trapping of gold as it has bedrock close to the surface and many boulders along its course. The area is ideal for camping, uncrowded, cool and in the pines. Stay off the creek for the first one-half mile in from the highway as it is a mineral withdrawal. I would not visit the area during the winter months as these mountains receive heavy snowfalls. The months of May, June, September and October are best; monsoon thunderstorms occur almost daily in July and August.

In Copper Basin, the land around Gabarina Hill is Phelps Dodge property. The hill is a mineralized intrusion now exhibiting the scars of heavy-duty mining from past years. The entire area surrounding Gabarina Hill, from the Sierra Prieta southward, is gold country. All the washes carry gold that is coarse, bright and coppery-tinted. Black sand is very abundant in your concentrate. You must be careful to stay off private property, as well as valid claims, when prospecting this district. Since it is lower in elevation than Copper Creek summertime temperatures are a little warmer and there is not as much shade. I also do not recommend visiting this area in the winter months or during the rainy season.

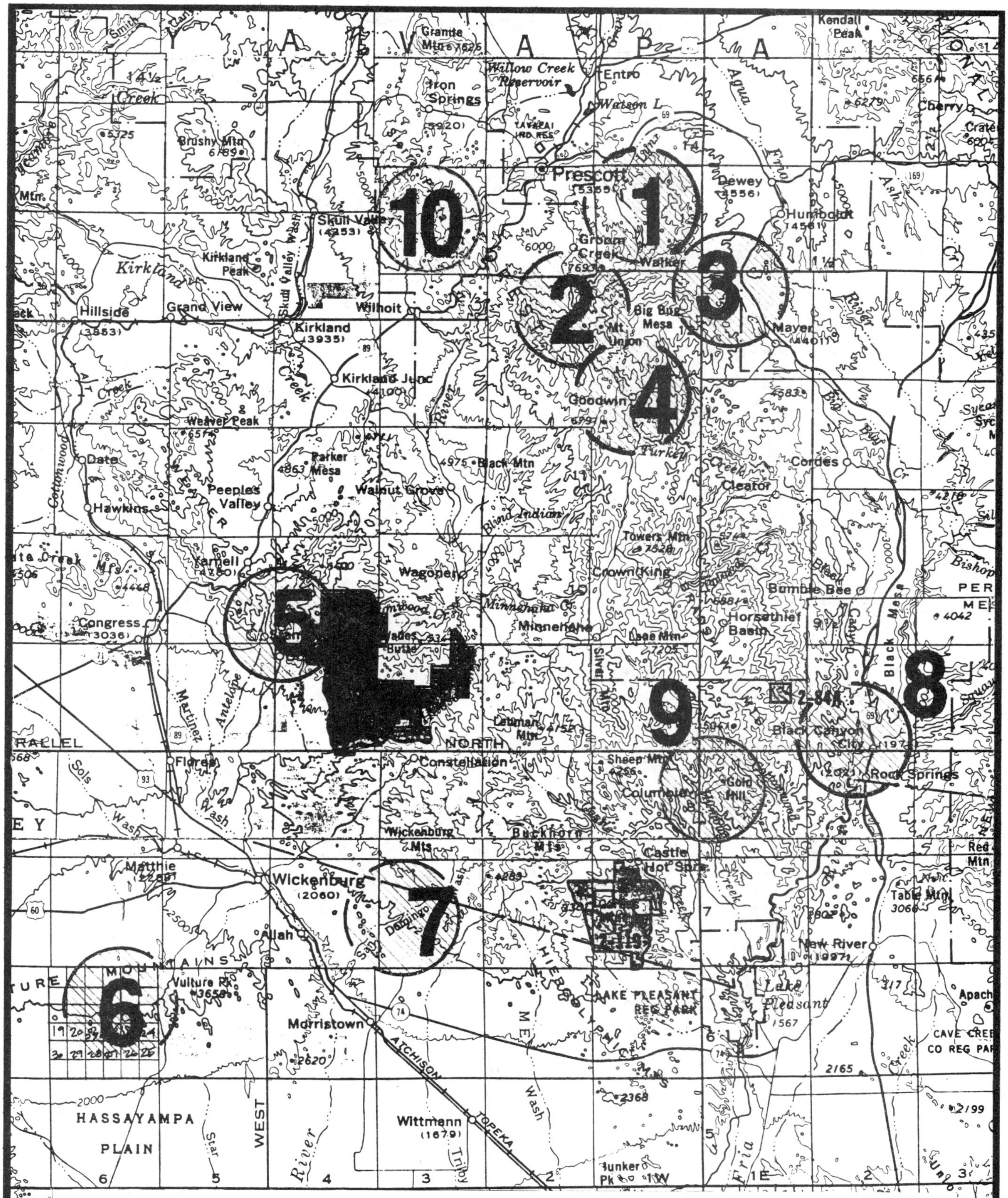

Fig. 24. **BRADSHAW MOUNTAINS, PRESCOTT AND VICINITY**
1, Lynx Creek; 2, Hassayampa Creek; 3, Big Bug Creek; 4, Turkey Creek; 5, Weaver Creek; 6, Vulture Mountains; 7, San Domingo Wash; 8, Soap Creek; 9, Humbug; 10, Copper Basin.

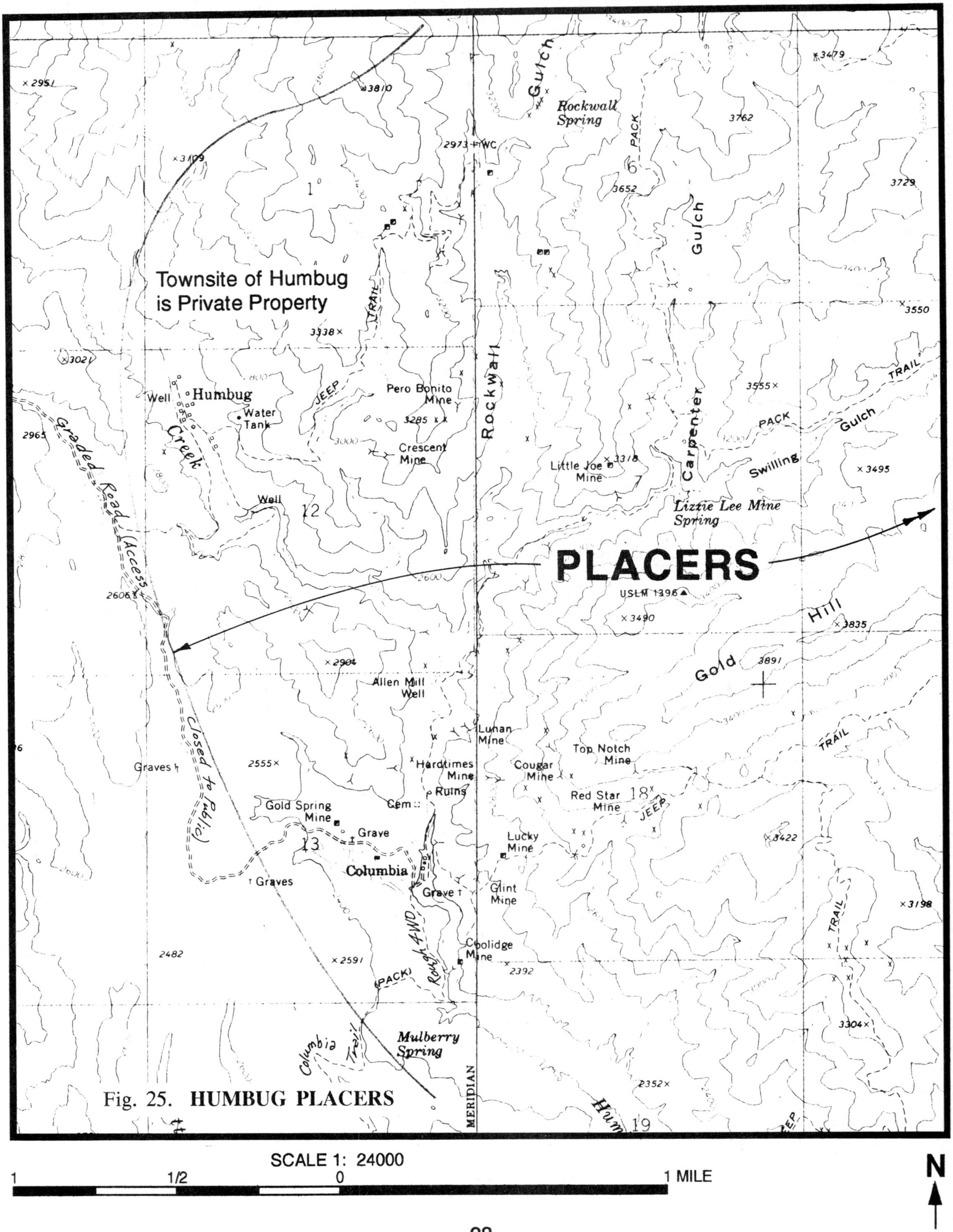

Fig. 25. **HUMBUG PLACERS**

Fig. 25a. On the Crown King Road north of Lake Pleasant, Yavapai County, Arizona, June 22, 1993, with the high Bradshaws in the distance. The 4WD pack trail to the ghost towns of Columbia and Humbug connects to the Crown King Road at the bottom of the hill (arrow). There are two abandoned homes here (Fig. 25b).

Fig. 25b. Two abandoned homes along the Crown King Road near the juncture of the Columbia Pack Trail. There is a third home, not in photo, occupied by a lone resident. Most of the land in this area is private but the Federal Government owns the mineral rights. The land around Columbia and Humbug is mostly BLM public domain.

Fig. 25c. Looking up Cow Creek, Humbug Mining District, Yavapai County, Arizona. I have panned the gravel of this creek, which has coarse gold accompanied by abundant black sand and garnets. There is a small flow throughout the year along some stretches at this locality. Bedrock exposures, suitable for sniping, abound along its length.

Fig. 25d. The start of the 4WD Columbia Pack Trail up the western flank of Black Hill, after crossing Cow Creek. Note washout in foreground from floodwaters of small wash cutting across road. This is a very rough road and challenging to skilled four-wheelers.

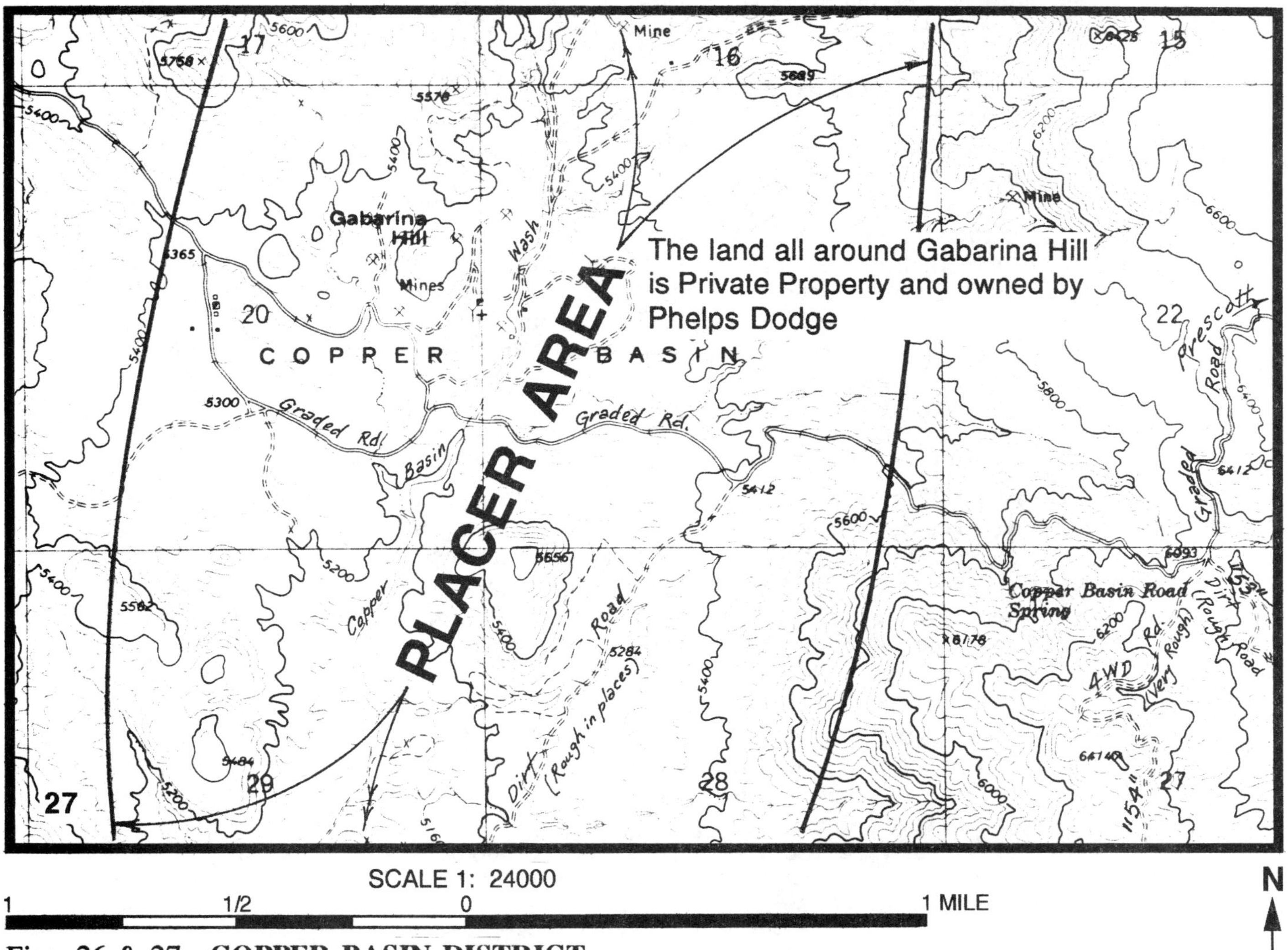

Figs. 26 & 27. COPPER BASIN DISTRICT
26, Copper Basin and Gabarina Hill in upper center, looking west, June 22, 1993; 27, Map of placer area.

WEST OF WICKENBURG (Figs. 28 & 29):

Most, if not all, travelers on U. S. Highway 60 - 70 between Wickenburg and Quartzsite are unaware of two gold placer fields easily reached by passenger car. Relatively close to each other, one is north of the highway and the other south of it. These are the **Harquahala Placers** in the **Little Harquahala Mountains** 10 miles south of **Salome**, and the **Desert Mine Placers** in the **Granite Wash Mountains** east of **Vicksburg** about 3 miles north of **Hope**. Summertime temperatures are very high here; be careful!

C. L. Shattuck discovered the Harquahala Placers in 1886 and worked these for about two years before the discovery of the Harquahala (Bonanza) and Golden Eagle lodes in 1888. The combined value of gold produced from both mines exceeds 50 million dollars based on a bullion value today of $400 per ounce. There is some mining continuing presently at both mines and some sporadic activity at other lodes in the area. Private property extends around the Rio Del Monte and Harquahala Mines (Figs. 28, 31 & 32). There are many claims throughout the area. A thorough search of the land status here in the La Paz County Recorder's Office, and the BLM office in Phoenix, is necessary to determine what ground is open to prospecting. The Little Harquahala Mountains are rugged and highly mineralized. There could be many opportunities for drywashing in the gulches away from the major mines and lode claims. Unfortunately, there are few dirt roads and tracks in the mountains and these are very rough.

You can reach the **Harquahala Placer Area** from **U. S. Highway 60 -70** in a passenger car. Go southeast on Navajo Street in **Salome** to **Harquahala Road** and take Harquahala Road south past the Indian Gardens residential development. The paved road ends just south of here at a barricade with a sign reading "Road Closed." A portion of the barricade was missing on the right-hand side. Tire tracks indicated that drivers ignored the sign and continued on the graded dirt road to the placer area. I also ignored the sign and drove on the excellent graded road to the placer area. The Harquahala Road continues southward from here and connects with Interstate I-10 (see Fig. 30). Time limitations, the welfare of my passenger (my wife), and the intense heat of the summer sun precluded me from doing a test in the area. At a more favorable time of the year, this would be an excellent place to test the gulches with a drywasher. Although the mountains are rugged and relatively remote you are not too far from civilization, and help, should you need it.

The **Desert Mine Placers** are very easy to get to. Go northwest from **Hope** on **State Highway 72** three miles to a graded dirt road on your right. This road crosses the railroad tracks. The paved road on your left is the Vicksburg Road. Ignore the dirt road with the unlocked gate and sign on it reading "Game Management Area 44A" that is also on your right just before reaching the graded road. This road parallels the railroad tracks and takes you away from the placer area. Heading northeast on the graded road, take the first dirt road on your right and follow this to a fork. Go northeast on the left-hand road about a mile to another fork. The right-hand fork takes you to the Desert Mine, a lode gold mine now inactive; the left hand fork crosses a large wash and continues on to the graded road. Just upstream of the wash crossing, I tested a small side gulch with my drywasher (Fig. 33). I recovered much fine gold and coarse, ragged grains from gravel on an inside bend of the gulch. There is a false bedrock of caliche-cemented gravel in places, and true bedrock exposures with a threading of quartz veins, in the gulch. The gravel consists of a flat, pebbly schist, quartzose rocks, and many small boulders. In places the gravel is only a thin layer on the bedrock, but on some of the inside bends it is more than two feet thick.

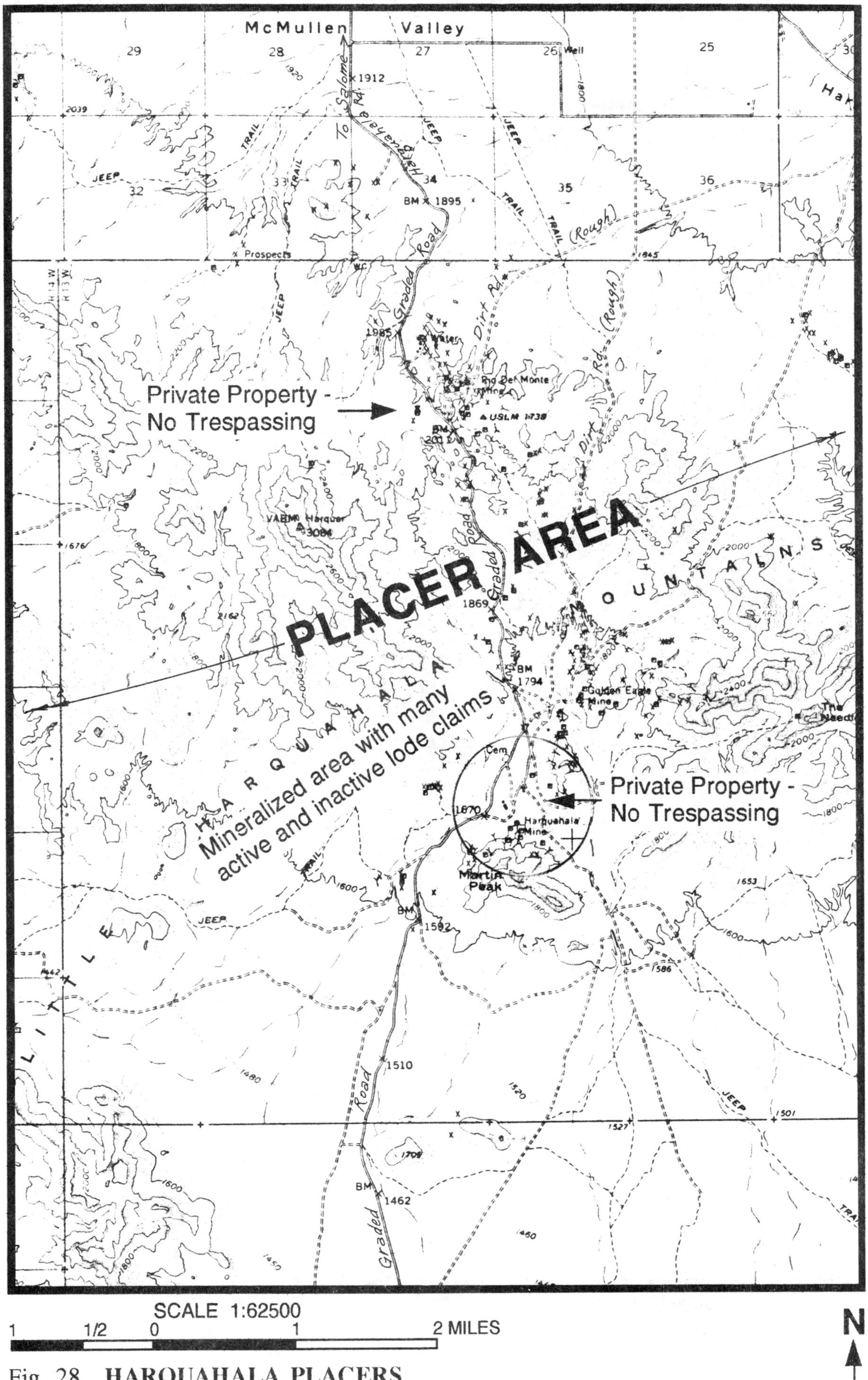

Fig. 28. **HARQUAHALA PLACERS**

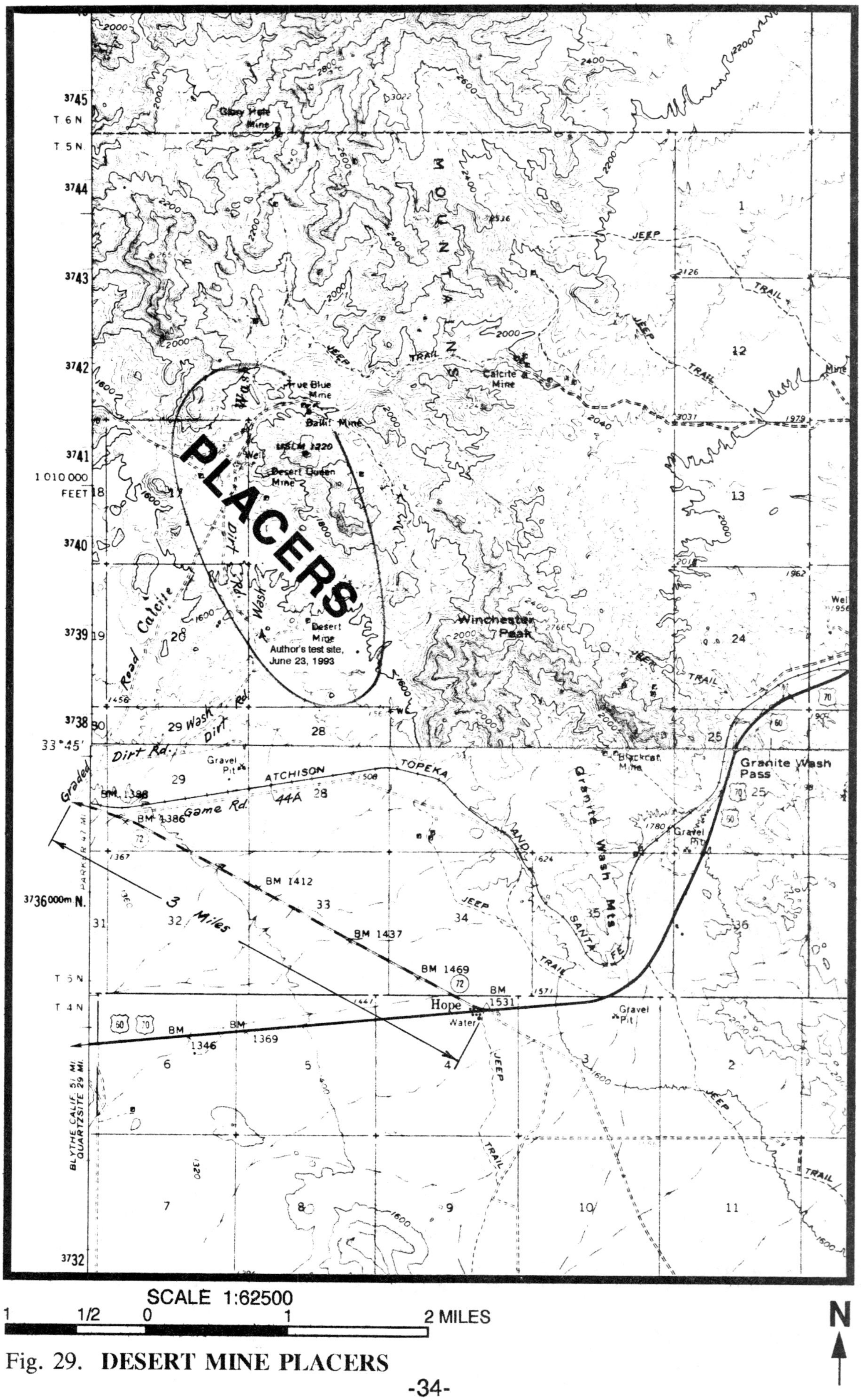

Fig. 29. **DESERT MINE PLACERS**

Fig. 30. The author's pickup truck parked in the "basin" just north of the Harquahala Mine property, June 23, 1993. In the background, on the low hill just visible above the cab, is the Golden Eagle Mine, a major past-producer of lode gold. The graded road in the foreground is the Harquahala Road from Salome to Interstate I-10.

Fig. 31. Looking south to Martin Peak and the Harquahala (Bonanza) Mine at its base, June 23, 1993. At the time of my visit there was some small development work going on at the mine. The land in the immediate foreground is not private property but there is a valid lode claim on it that is well-monumented.

Fig. 32. Harquahala Mine property at the base of Martin Peak, June 23, 1993. The building on the low hill is vacant and in utter disrepair. Note the "No Trespassing" sign in the foreground proclaiming this ground as private property.

Fig. 33. **DESERT MINE PLACERS NEAR VICKSBURG.** Author screening gravel from a small wash preparatory to drywashing, June 23, 1993. Abundant coarse, pin-head size gold occurs here in gravel composed mainly of slabby pebbles of schist and quartzose rock. Air temperature at time of photograph about 103 degrees F.

EAST OF QUARTZSITE (Fig. 34):

The area around Quartzsite is classic gold country and it attracts winter visitors from all over the United States. Most come for the weather and the comradeship; but, each year a growing number of gold-seekers come armed with metal detectors. How much gold they find is anyone's guess. When the hot summer sets in most have left, leaving behind a permanent contingent of die-hard gold-seekers and others who just call Quartzsite home.

It is no secret that the desert around Quartzsite has gold. Spectacular finds of metal-detected nuggets from here receive much attention in the press, particularly in treasure magazines. The surprising fact is that much of this gold is practically on the surface just waiting for the lucky person with the right metal detector. What are your chances of finding this gold? Although luck plays an important part, your chances increase using the proper equipment and the amount of time you spend using it, particularly so when metal-detecting. Don't expect to hit a nugget right-off-the-bat; be prepared to put in some time. Most of all, be patient. This is good gold country and it's just a matter of time if you keep at it and don't get discouraged.

I reported on the La Paz, Oro Fino and La Cholla Placers in ***Arizona's Golden Secret***. Here is another place you can go to, and find, gold. Don't forget to bring along your metal detector, or a drywasher. I usually bring both.

The **Plomosa Placers** in the **Plomosa Mountains** are about seven miles directly east-southeast of Quartzsite on the western slopes of the mountains (Figs. 35, 36 & 37). You can reach the drywashing placer area on the **frontage road** that connects with **Exit 19 south of Interstate Highway I-10**. Head east on the paved road signed "Dead End." At the first right-angle bend the road becomes a good graded dirt road that goes all the way to the diggings, a distance of about two miles. It is suitable for passenger cars. At the end of the straight section you come upon a mobile home, and private property, on your right (Fig. 38). A short distance beyond, the road forks. The left-hand fork takes you to an active placer mining operation (Fig. 39). This is near the site of the now-vanished mining community of Plomosa and it is private land. The right-hand fork continues on to another patented placer. However, north and south of these, about a half mile away, is public domain land, and if not already under claim, open for prospecting. You can reach the public domain land north of the private property by taking any of the dirt tracks leading off the dirt road you came in on. Most of the land around is public domain land, but you should try to get as close to the source of the gold as you can. This means getting close to the mountain flanks.

I have drywashed **Smith Wash** (Fig. 37), just west of the first mobile home, on public domain land. This is a "braided wash." What this means is that the wash has many channels that diverge and merge along its course. Like most of the washes issuing from these mountains, Smith Wash's gravel contains billions of small rocks of rather uniform size. There is very little "dirt" in this gravel. Within one to two feet below the surface of the wash the gravel becomes firmly cemented, requiring the use of a steel digging bar. This condition is typical of a "desert placer" whose washes see little in the way of running water except in times of flood. Bedrock is very deep. When mining these washes, I select spots just downstream of a juncture on the "quiet" side, an inside bend, or downstream of an obstruction. By studying the way the gravel deposits you can determine the quiet side where the gold drops out. Pay particular attention to the gravel bars just downstream from bushes growing in the bed of the wash. By digging into the gravel just behind the bush you should get the gold tangled up in the root system. In places, the high bank along Smith Wash indicates mining done during the 1930s; the remains of tailings piles and rusty cans of the lead-seal type are still in evidence. I had very little luck mining the high bank here.

The gold I found in Smith Wash is small, ragged and with a distinctive coppery tint. Its source is near at hand. Mercury coats some of it. This may be the result of the early miners' use of mercury to catch the fine gold, with much mercury being lost, ending up in the washes, and apparently coating some of the gold there.

Generally, it will pay you to test the small washes with a drywasher, particularly near the diggings. If you brought along a metal detector you should detect not only the gulches but also the flat mesas between them (Fig. 40). The gold, eroding out of many gold-bearing veins and stringers in the mountains for millions of years, has spread out over many square miles. The miners of old would have had to process the entire area, including the mountains, through their gold recovery equipment to get all the gold, but they couldn't and didn't. They worked only the richest portions and still left behind much gold. The processes of erosion, mostly rainfall, and to a lesser extent, wind, constantly expose this gold. Every year, after the winter rains, a new crop of gold exposes on the surface. The knowledgeable winter visitor, who comes here for more than just the pleasant winter climate, is sure to include a metal detector in his baggage for just this situation.

The Plomosa Mountains are highly mineralized. They, with their gold, extend well north of the Interstate all the way to Bouse. There are several patented lode gold mines in these mountains, as well as unpatented placer and lode gold claims. Although claim stakes abound, many of the claims are invalid from lack of assessment work. However, you should always check out, beforehand, the land status of the area in which you plan to prospect to avoid any misunderstanding with private landowners, or claimants. If being on someone else's claim concerns you, placer claims are available on a daily fee basis. Inquire at the tourist information center in the **Quartzsite Chamber of Commerce** on north Highway 95, or you may write to the chamber at **P. O. Box 85, Quartzsite, Arizona 85346, Phone 1-(602) 927-5600.**

Northeast of Quartzsite, just off the **Plomosa Road** (Quartzsite-Bouse Road), is the **Plomosa (North) Mining District** notable for copper-gold and manganese deposits. This is lode mining country, but few know that the washes in this district carry placer gold that is small, ragged and of a coppery color. I have drywashed the small washes draining northward from the **Mudersbach Mine** area (Fig. 41) with excellent results. These washes are the typical "desert placer" kind. Use the same approach to mining this area as you would other placer districts around Quartzsite. To reach this locality, drive north from Quartzsite on **State Highway 95** a distance of about six miles to a road on your right signed "Plomosa Rd." Take this paved road northeastward a distance of about 10 miles to the mineralized area. The dirt road to the Mudersbach Mine is on your right a mile or so beyond. You can negotiate it in a passenger car. The Southern Cross Mine shown on the map is private property and not available for prospecting.

A final note: The placers of the Quartzsite area are ideal for drywashing. This part of Arizona receives very little rain except for the occasional winter storm. The months of June, July and August are very hot. Ironically, it is during these months that the drywashing gravel here is at its best - bone-dry - but very dusty; and, there's no competition. Even the local gold-diggers lie low during the heat of the summer.

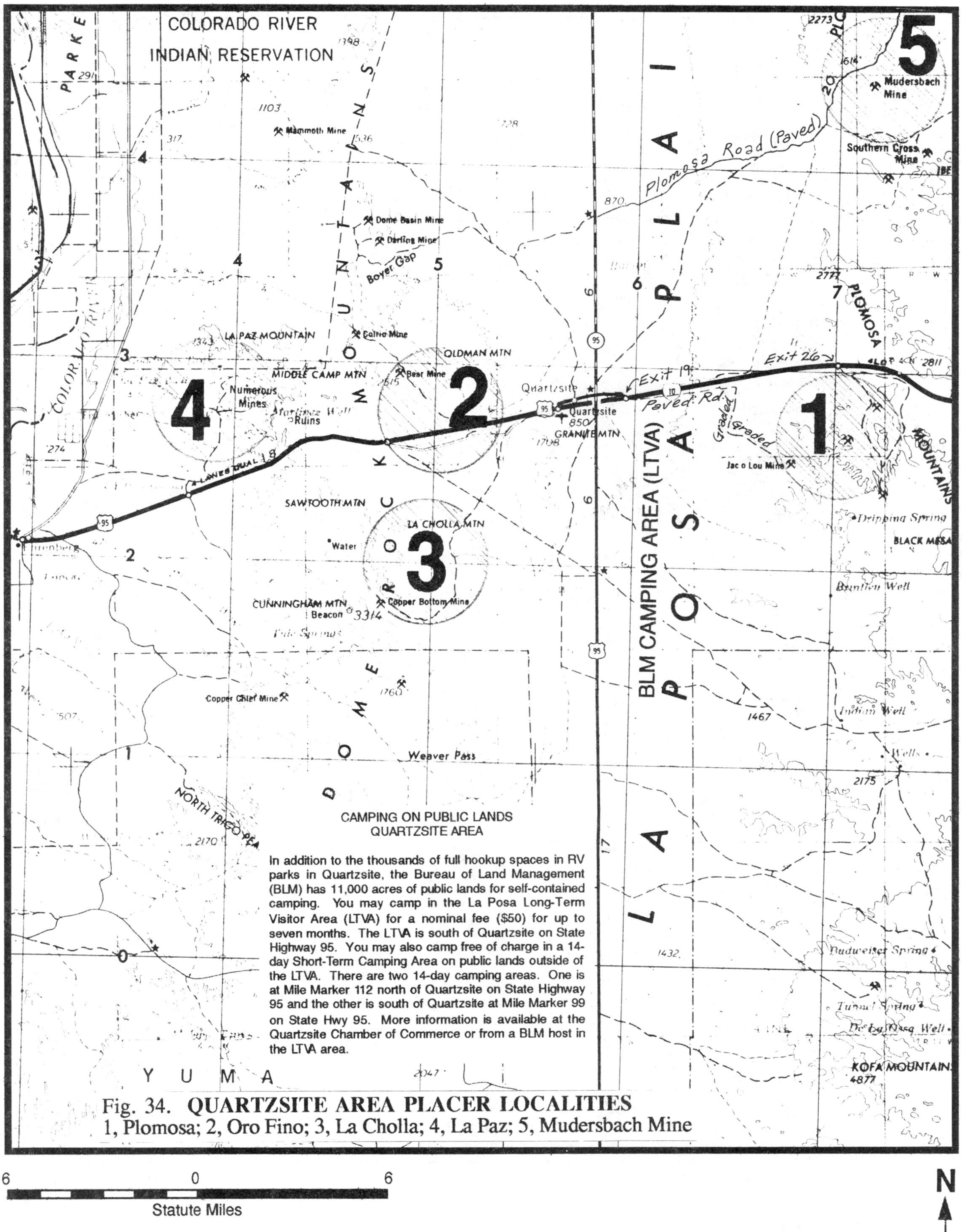

CAMPING ON PUBLIC LANDS
QUARTZSITE AREA

In addition to the thousands of full hookup spaces in RV parks in Quartzsite, the Bureau of Land Management (BLM) has 11,000 acres of public lands for self-contained camping. You may camp in the La Posa Long-Term Visitor Area (LTVA) for a nominal fee ($50) for up to seven months. The LTVA is south of Quartzsite on State Highway 95. You may also camp free of charge in a 14-day Short-Term Camping Area on public lands outside of the LTVA. There are two 14-day camping areas. One is at Mile Marker 112 north of Quartzsite on State Highway 95 and the other is south of Quartzsite at Mile Marker 99 on State Hwy 95. More information is available at the Quartzsite Chamber of Commerce or from a BLM host in the LTVA area.

Fig. 34. **QUARTZSITE AREA PLACER LOCALITIES**
1, Plomosa; 2, Oro Fino; 3, La Cholla; 4, La Paz; 5, Mudersbach Mine

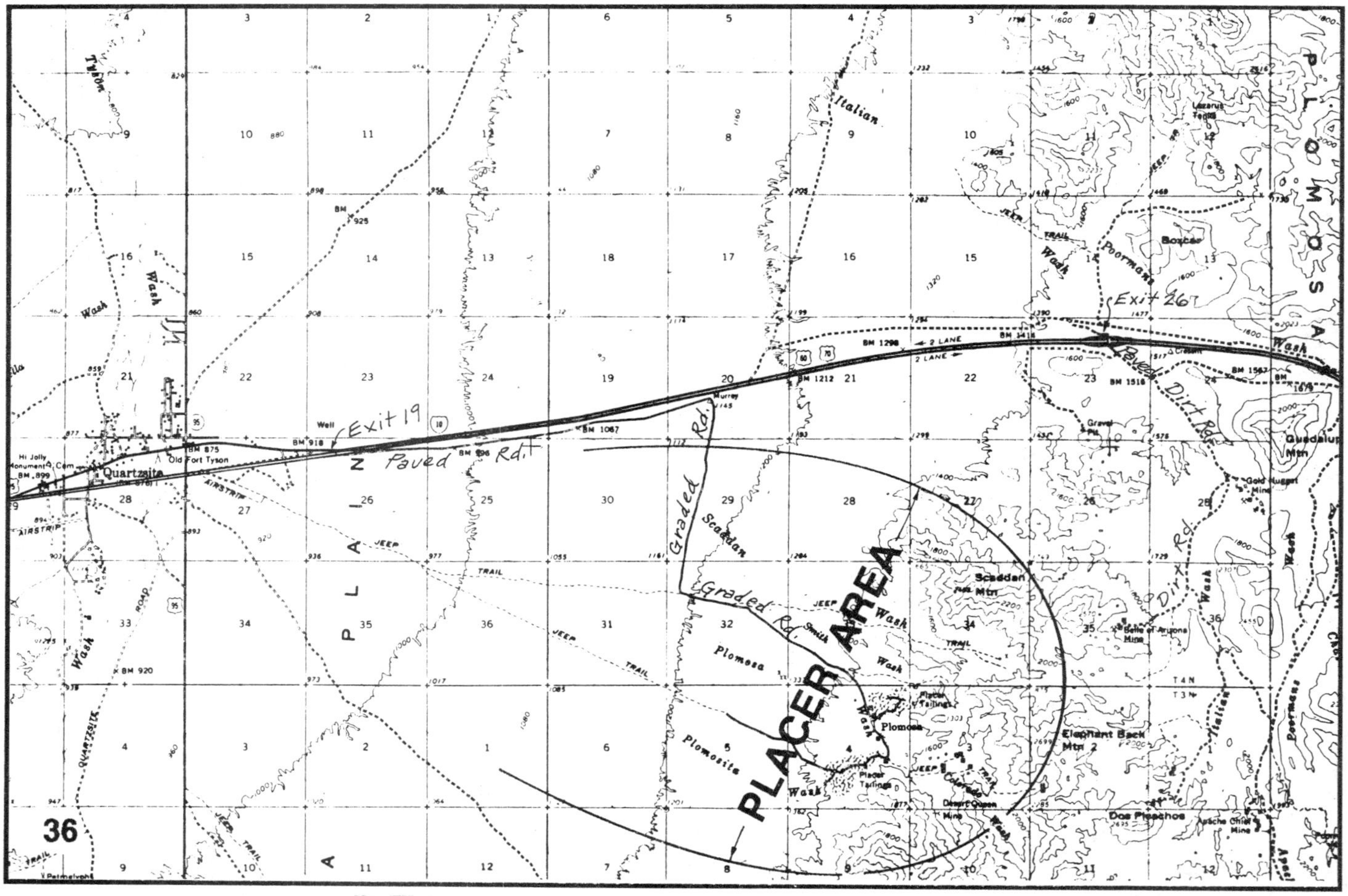

SCALE : Statute Miles
2 1 1/2 0 1 2

N

Figs. 35 & 36. PLOMOSA PLACERS
35, Author's pickup truck on the graded road to the placers, July 20, 1993; 36, Map of placer area.

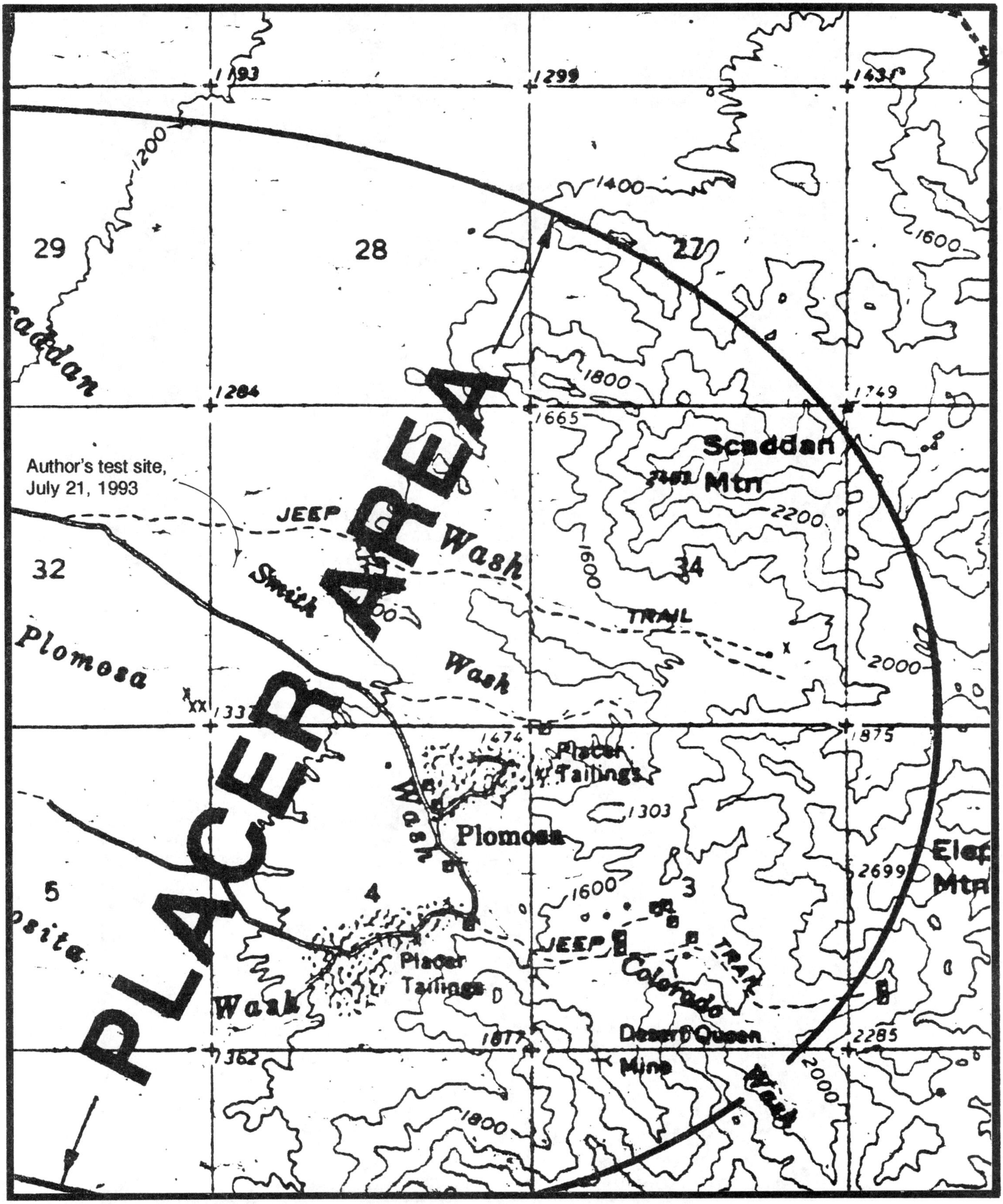

Fig. 37. **PLOMOSA PLACERS - ENLARGED MAP**

N

Fig. 38. Mobile home and beginning of private land near the site of the now-vanished community of Plomosa. Plomosa mountains in the background. July 20, 1993. I drywashed Smith Wash, which is to the left of the photo, downstream from here and off private land.

Fig. 39. Active placer mining operation near the site of the now-vanished community of Plomosa. Note the tailings piles at left center of photo; Scaddan Mountain in the background. This operation is on private land. July 20, 1993.

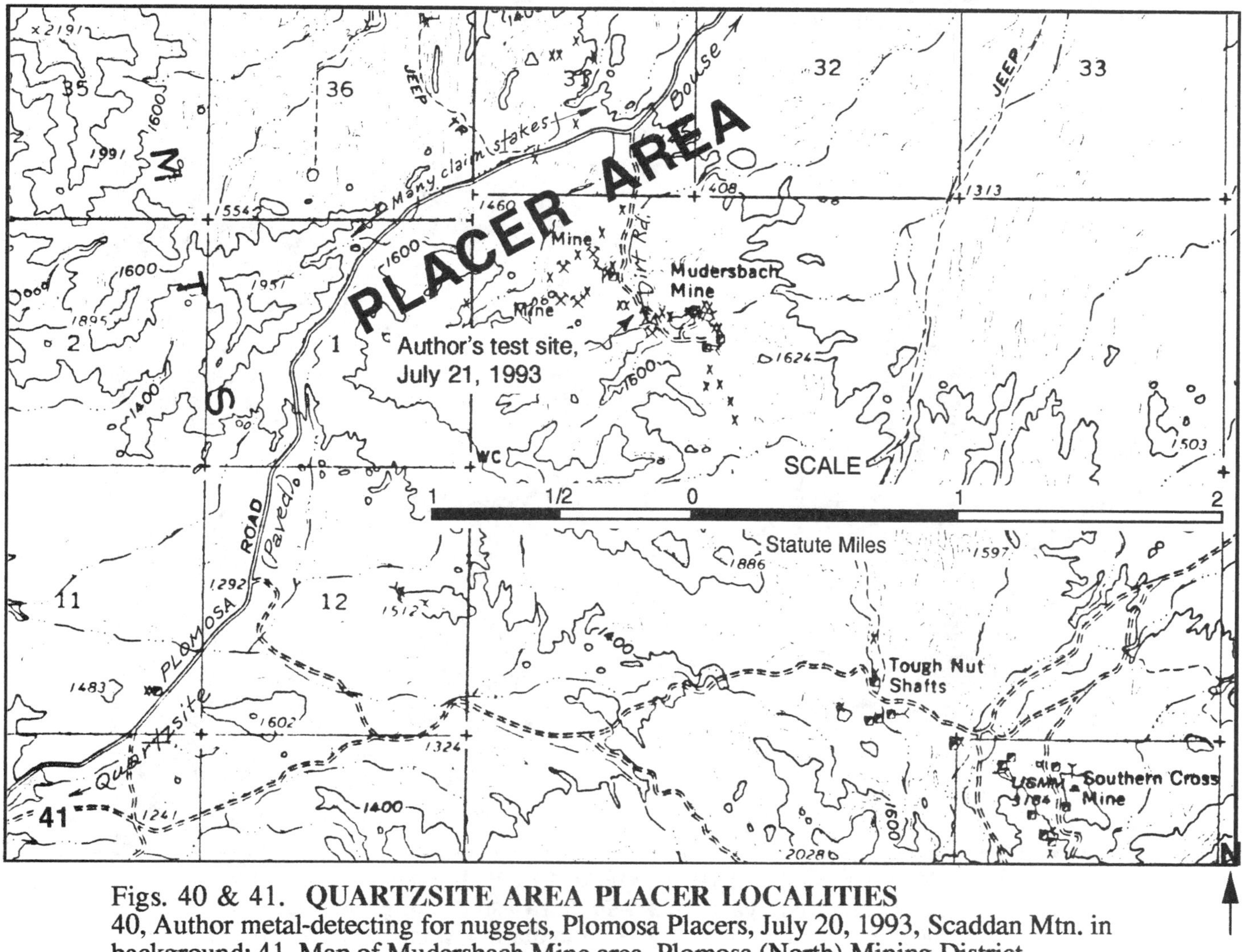

Figs. 40 & 41. **QUARTZSITE AREA PLACER LOCALITIES**
40, Author metal-detecting for nuggets, Plomosa Placers, July 20, 1993, Scaddan Mtn. in background; 41, Map of Mudersbach Mine area, Plomosa (North) Mining District.

Fig. 41a. **PLOMOSA (NORTH) MINING DISTRICT - MUDERSBACH MINE AREA.** Looking to the southeast toward the mines located in upper center of the photo, September 19, 1993. The washes draining away from the mines and prospects carry fine, coppery-tinted gold.

Fig. 42a. **PLOMOSA (NORTH) MINING DISTRICT - MUDERSBACH MINE AREA.** Author's drywasher and vacuum set up in a wash draining from one of the lode mines. I have processed the gravel from the excavation and am now vacuuming up all the remaining gravel from the many cracks, fissures and pockets in the bedrock.

NORTH OF KINGMAN (Fig. 42):

In the far northwestern part of Arizona, near the **Colorado River**, are two placer fields relatively close to each other. One is the **Gold Basin District**; the other is the **King Tut**, or **Lost Basin, District**, and both ideally suited for drywashing and metal detecting. A prospector named Dunlop discovered each in the 1930s, a rather recent date as far as placer discoveries go. The interesting thing is that lode gold mines had been operating in both districts in the late 1880s, well before the placer discoveries. Usually, prospectors of old were very astute at finding gold. For them not to have found these placers before, or after, the lode discoveries, is indeed surprising.

During the depression years of the 1930s, hundreds of unemployed men in both districts worked the gravel with drywashers. Their machines were not very efficient and much gold went off their screens or lost in the tailings. The drywashing piles left behind from their activities offer many opportunities for metal detecting.

You can reach both districts by driving northwest out of **Kingman** on paved **U. S. Highway 93**. Motorists, going to and from Las Vegas, Nevada, give it very heavy use. About 28 miles northwest of the intersection of I-40 and U. S. 93 a paved road on your right connects with the highway and leads northeast to **Dolan Springs**. This is the **Pearce Ferry Road** that goes all the way to the Colorado River. Dolan Springs is a small community composed mainly of retirees living in mobile homes. It was the subject of much negative media attention in the early 1970s. In this corner of Mohave County are several real estate developments that were nothing more than scams. They gave Arizona real estate a black eye. As a result, Arizona's real estate laws were overhauled and enforced, which effectively stopped this abuse. The scars of roads cut into the desert during the 1950s and 60s for these purported developments are noticeable as you drive along Highway 93 and the Pearce Ferry Road to Meadview.

Gold Basin is your first stop along the Pearce Ferry Road (Figs. 44 & 45). Approximately 18.1 miles out from U. S. 93 is a good dirt road on your left heading northwest toward the mountains. A passenger car can negotiate it. There is a stop sign on this road at the turnoff. Take this road and drive about four miles to the fork. The left-hand fork goes to the OK Mine and the right-hand one to the Malco and Excelsior mines. On the hill to your right is the Excelsior Mine (Fig. 47). All these mines were lode gold producers. The washes draining away from the mines here carry gold that is small, ragged grains, bright brassy-yellow in color, with some attached, light-brownish oxide. The small wash that I tested contained gravel that was very silty and sandy, with very few large rocks (Fig. 46). It became more compact with depth. Although I excavated more than two feet into the gravel near a bedrock outcropping, I did not reach bedrock nor the compacted false bedrock so commonplace in desert placers.

The desert around the mines contains the refuse of early-day miners, mostly rusty cans from the days of the Great Depression. In places, green bottle glass revealed itself, with broken chinaware here and there. In the late 1870s and 1880s, prospectors discovered the lode gold mines here such as the O. K., Malco and Excelsior. Opportunities for serious artifact collecting and metal-detecting are many in this area. Because the ore from the early operations was free-milling, the dumps may hold hidden treasures discoverable only with a metal detector.

Not far from here is the ghost mining camp of Cyclopic. The Cyclopic mine, discovered in the 1880s, was a major producer of lode gold. Cyanide leaching was the process used on much of the gold ore from the mines. You can still see the remains of vats and waste tailings from the leaching operations at the Excelsior and other mines.

King Tut, or **Lost Basin**, is approximately 12.9 miles farther on from the turnoff to Gold Basin (Figs. 48 & 49). Stay on the Pearce Ferry Road until you come upon real estate developments of dirt roads and scattered mobile homes. From here westward, to the foot of the mountains, is the King Tut Placers. This is mostly private land except for a thin strip of public domain, which starts at the foot of the mountains and extends westward to the boundary of the Lake Mead National Recreation Area. Section 4, T 30 N, is for the most part public domain land, as is Section 33, T 29 N (see Figs. 48 & 49). There are some claims in the area.

The good dirt road to the King Tut Mine area (Fig. 50) is on your left as you head north on the Pearce Ferry Road, approximately 12.9 miles from the Gold Basin road turnoff. About a mile in from the highway you come upon an intersection with another dirt road. Go left on the dirt road and bear to your right. You come upon an adobe ruin in a wide, flat area. A small wash is on your right and the tailings of modern-day drywashing are noticeable. In the distance to the West is a huge tailings pile at the King Tut Mine.

I tested the small wash with my drywasher. A firmly cemented false bedrock of coarse rocks and gravel occurs not more than a few inches below the surface of the wash in some places, and is as much as one to two feet deep in others. The bedrock is very rough and pocked, but I did not come across any cracks or fissures. The gravel carrying the gold is very silty and contains many small pebbles firmly compacted in places. Because the gravel varied in thickness, and paystreaks are usually erratic in desert placers, I put through my drywasher the entire thickness of gravel, from surface to bedrock. Upon reaching bedrock, I did what I would normally do with any bedrock. **I swept it clean!**

The gold I found here is mostly flour and pinhead size, but I did recover particles up to 1/20 inch in diameter. It is very coarse and ragged, bright brassy-yellow in color, and has much reddish-brown oxide attached. Most gold came out of the layer directly on bedrock and from the small pockets in the bedrock, but virtually the entire gravel layer contains gold. If you decide to mine these upper washes near the mines, my advice to you is to process the entire gravel layer through your gold recovery equipment. Save all of your black sand concentrate for further processing due to the abundance of flour gold.

The Gold Basin and King Tut placers are ideal for drywashing. Although the summer is hot, if you can tolerate the heat, it is a good time to visit the areas. You will have the places to yourself. This part of Arizona receives very little rain except in the winter months. So, for most of the time, the gravel is bone dry and amenable to drywashing. Even where firmly compacted, the gravel breaks up readily with a shovel and digging bar. However, the firmly cemented false bedrock of small rocks, such as at King Tut, does not break up readily even with heavy blows of the digging bar. I recommend that you do not expend precious time and energy trying to break it up. Gold is plentiful enough in the gravel resting on the bedrock.

In the King Tut placer area are exposed high banks showing millions upon millions of small pebbles rather evenly distributed throughout. It might pay to work these high banks but it will require much effort on your part. These banks are hard as concrete making excavation an extremely strenuous venture, unless...you are in it for the exercise.

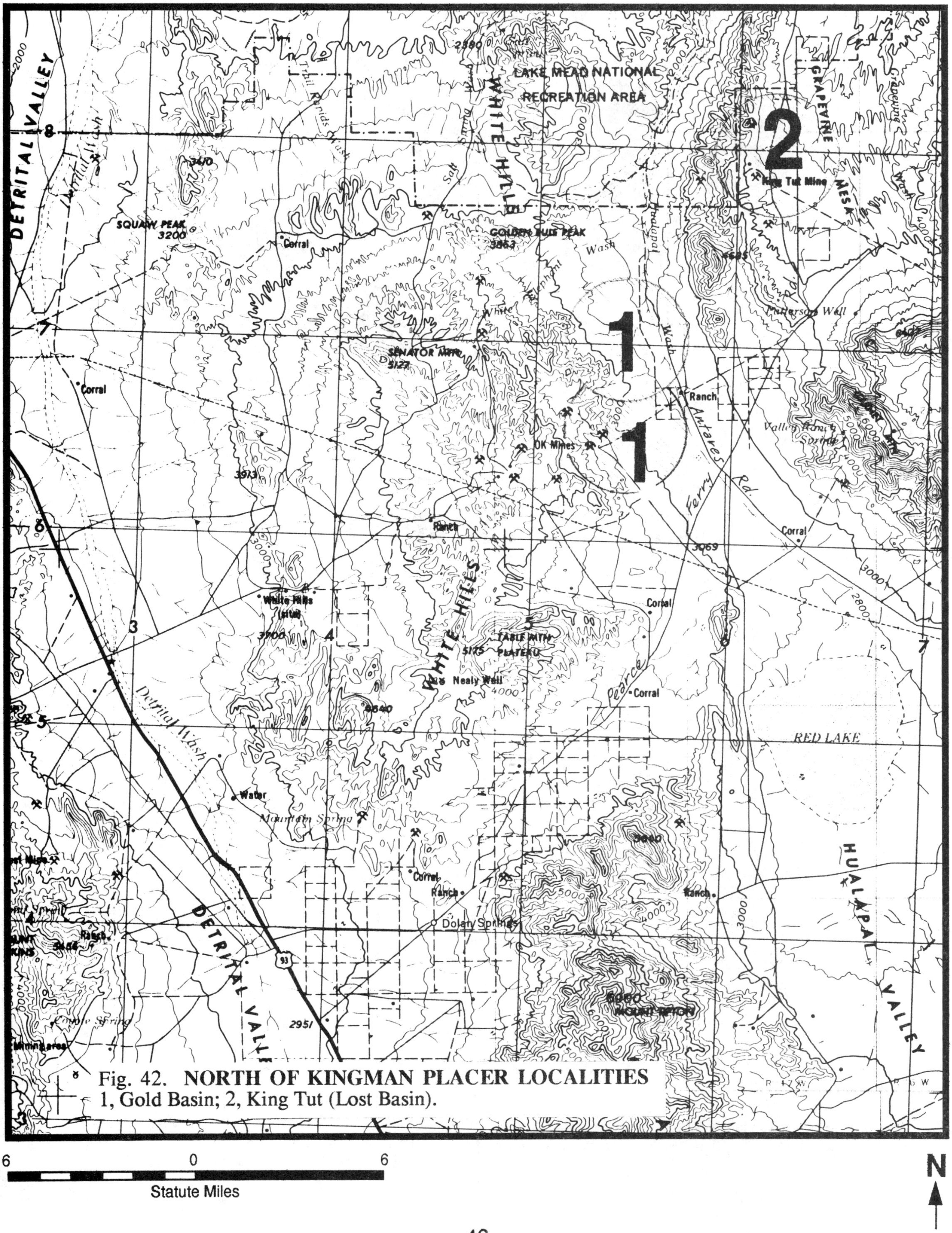

Fig. 42. **NORTH OF KINGMAN PLACER LOCALITIES**
1, Gold Basin; 2, King Tut (Lost Basin).

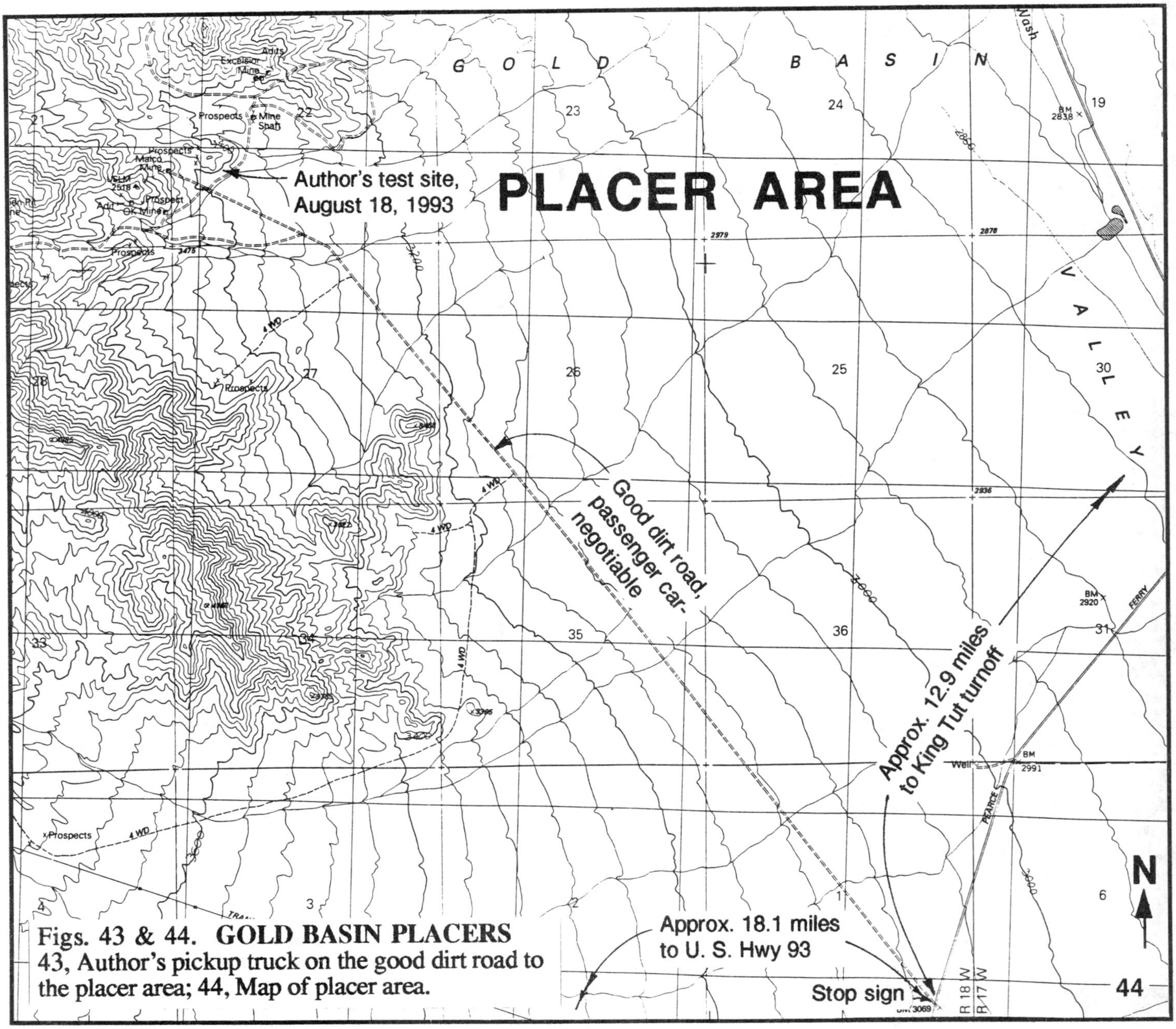

Figs. 43 & 44. **GOLD BASIN PLACERS**
43, Author's pickup truck on the good dirt road to the placer area; 44, Map of placer area.

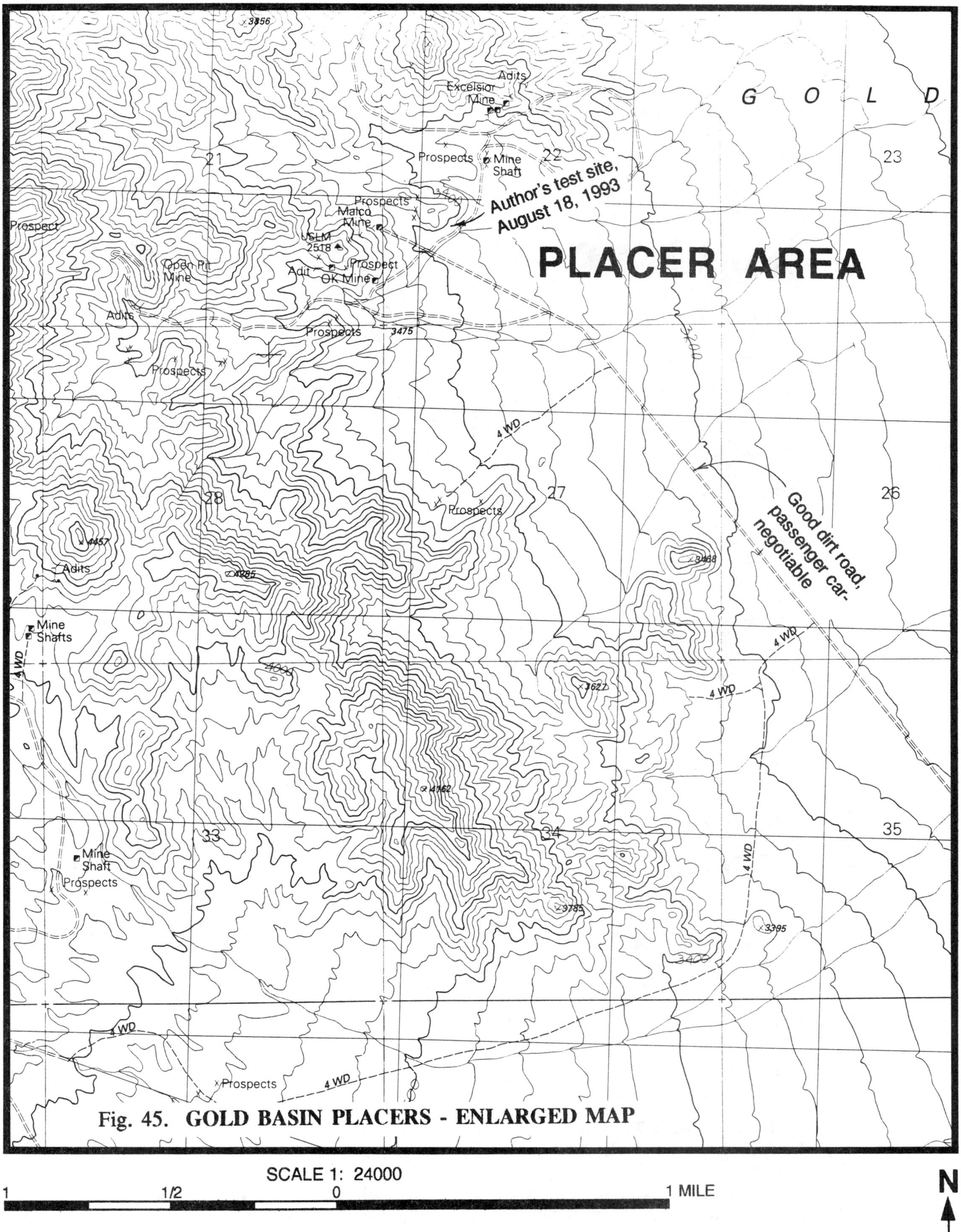

Fig. 45. GOLD BASIN PLACERS - ENLARGED MAP

Fig. 46. **GOLD BASIN PLACERS.** Author testing a small wash near the Malco Mine. I am shown here shoveling the gravel onto the grizzly screen attached to the hopper of my drywasher. The gravel is very silty and sandy, becoming more compact with depth, and contains only a few large rocks. The gold I found is small, ragged grains, bright brassy-yellow in color, with some attached, light-brownish, oxide.

Fig. 47. **GOLD BASIN PLACERS.** The remains of the Excelsior Mine. In the foreground are the ruins of the ore processing plant. On the hill in the upper center of the photo is the disintegrating headframe above the inclined shaft, now mostly caved. Behind the headframe is an adit with a perilous roof and some shoring inside. Stay out!

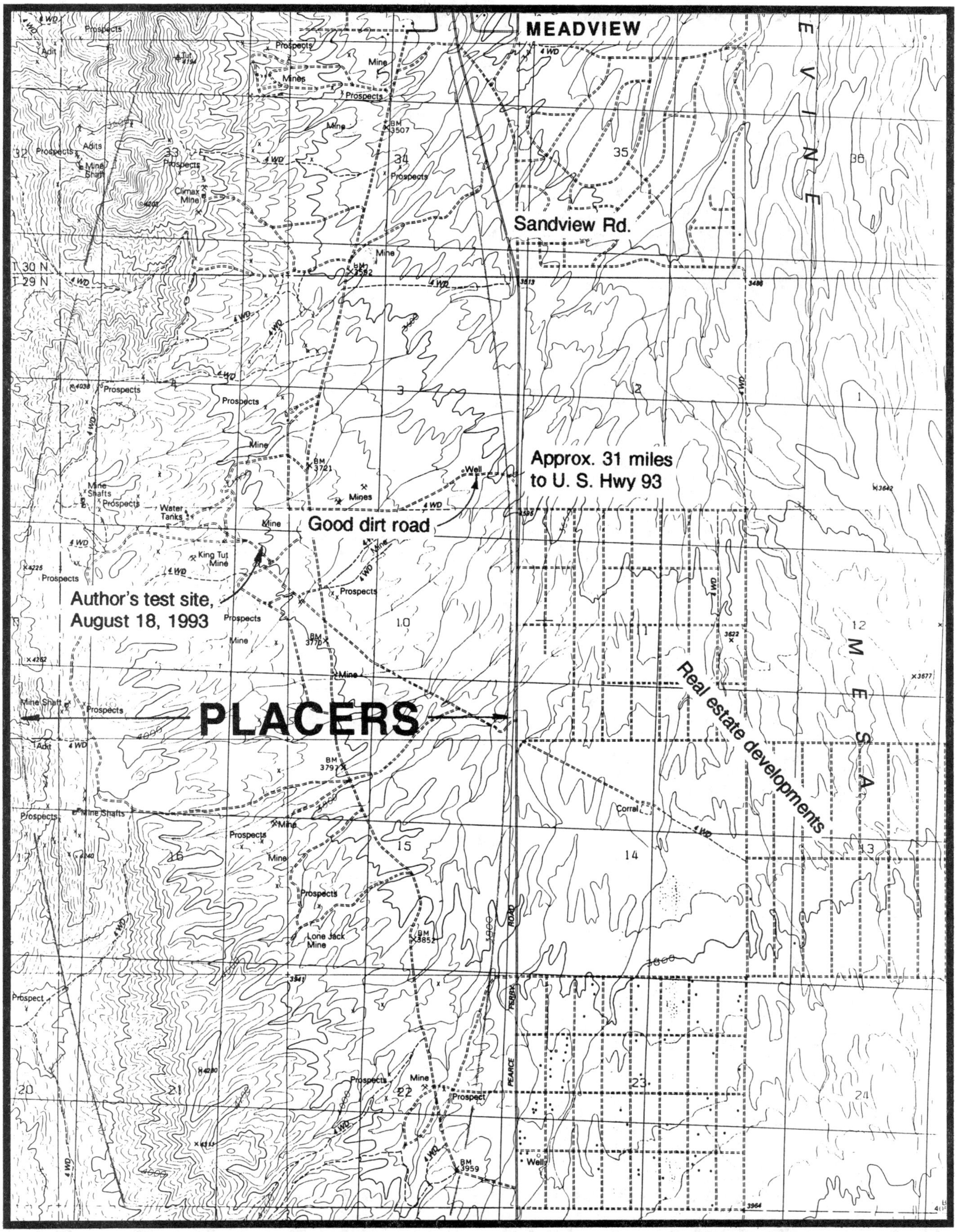

SCALE : Statute Miles
1 1/2 0 1 MILE

N

Fig. 48. **KING TUT PLACERS**

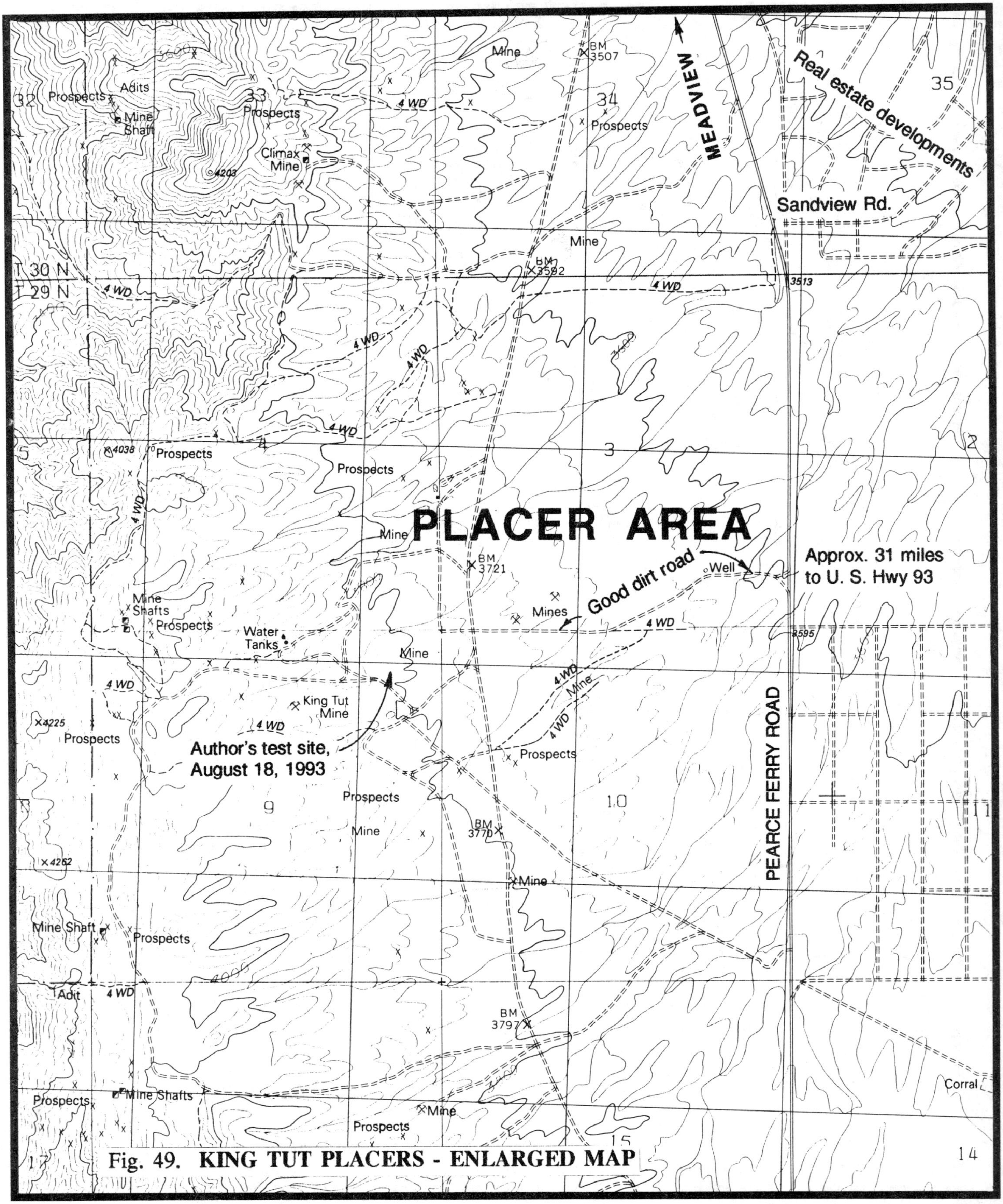

Fig. 49. KING TUT PLACERS - ENLARGED MAP

SCALE 1: 24000

1 1/2 0 1 MILE

N

Fig. 50. **KING TUT PLACERS.** Typical good dirt road heading west to the placers. This road is passenger car-negotiable. The major placer area is on the pediment at the base of the low hills in the distance, and extends beyond the edges of this photograph. The King Tut Mine is at the extreme upper left of the photograph.

Fig. 51. **KING TUT PLACERS.** The author's vehicle parked at the base of a Joshua Tree at the King Tut Mine area. To the extreme right of the Joshua Tree are the ruins of an adobe building, and in the distance are the Grand Wash Cliffs. I drywashed a small wash here, which is to the left of the truck and out of the photo, and got much fine gold.

Fig. 52. **KING TUT PLACERS.** The adobe ruins of a rather large mine building, its past purpose unknown to me. In the right center of the photograph, beyond the ruins, is a huge tailings pile from the King Tut Mine. Large bushes now growing on it indicate that this pile is very old. Metal-detecting opportunities abound here.

Fig. 53. **KING TUT PLACERS.** The Grand Wash Cliffs in the background. In the immediate foreground are the tailings piles left by modern-day drywashers, probably of the gasoline-powered blower type. The area is littered with modern-day items, including plastic gasoline cans, indicating drywashing by others using motorized equipment.

YUMA AREA (Fig. 54):

There are several small placer fields close to Yuma that are fairly easy to reach in a passenger car, although you may have to do some hiking if you don't have a four-wheel-drive, or high-clearance, vehicle. I reported on one of these, the Gila City Placers, in ***Arizona's Golden Secret***. A word of caution to those contemplating prospecting the mountain ranges around Yuma. The terrain is rugged and summertime temperatures are very high. **I do not recommend any prospecting in the Yuma area, or any other area of Arizona lower than 3,000 feet in elevation, during the months of June, July and August.** If you go ahead anyway and prospect in spite of this advice, take along plenty of drinking water. Figure on a **minimum** of two gallons per person per day.

Our government has withdrawn from mineral location vast tracts of Yuma County, and neighboring La Paz County, containing *significant* mineral resources, by the establishment of military reservations, wildlife refuges and Indian reservations. I intended to report on several placers now locked up within the boundaries of these reservations, particularly the Kofa Game Refuge (see Figs. 70, 71, 77 & 78). Unfortunately, the law pertaining to the prospecting for, and removal of (**no matter how small the amount**) minerals from withdrawn areas, is very clear. A publication by the U. S. Department of the Interior cites the following.

> **"Individuals who disturb resources after the effective date of withdrawal and who do not have valid existing rights may be considered in trespass and can be held liable for trespass damages. In addition, trespassers may be fined and sentenced to a term in jail."**

Additionally, our government has set aside as wilderness large portions of Yuma, La Paz, and other counties of Arizona containing significant mineral resources. Although the minerals in these wilderness areas are not totally withdrawn, they may as well have been. Very stringent regulations apply to the mining of minerals in wilderness areas, including the prohibitions against the use of any motorized equipment and the locating of mining claims.

The **Las Flores Placers**, on the southeastern flanks of the **Laguna Mountains**, are directly across the Gila River from the Gila City Placers (Figs. 55 & 56). Prospectors discovered gold in the area before 1865. Mexican and Indian placer miners founded the town of Las Flores sometime in the 1860s, nothing of which remains today. Several inactive lode gold mines scatter about on the lower flanks of the mountains. Numerous, small, gold-bearing quartz veins and stringers occur in the schist and probably were responsible for the placer gold found in the washes and gulches.

You can reach the placers by heading east from Yuma on **U. S. Highway 95**, then going north across the **McPhaul Bridge** over the Gila River. After the highway makes the right-angle turn and heads northwards, it parallels the mountains and the placers. The boundaries of the Yuma Proving Ground lying to the West and North, with the highway on the East and South, squeeze down this placer district to one of the smallest in Arizona. A four-wheel-drive dirt road going into the placers leads off the highway 1.1

miles from the northern foot of McPhaul Bridge (Fig. 55). Unfortunately, this road is washed-out a hundred yards in from the highway, and in many other places along its length. I entered the area on a dirt access road 1.8 miles from the northern foot of the bridge and followed a dirt track that went into a dry wash. I continued on, heading northward to another wash, and parked in a wide, level area. From here I prospected on foot. You, too, can prospect on foot the many washes and side gulches draining off the mountains, but pay close attention to the military reservation shown on the map, which is off-limits to mining. I do not advise driving on the power line maintenance roads as this is illegal. It may tempt you to access the area from the highway through the flats that parallel it. This is risky. There is a deep layer of silt here that is fairly solid in the dry season, but if there is rain, watch out! With the slightest wetting it becomes a quagmire, and without a four-wheel-drive vehicle, I would almost guarantee your getting stuck.

I prospected a small wash west of where I parked. Along one side of the wash is evidence of high-bank mining from past years. My first test was behind a bedrock outcrop extending into the wash (Fig. 58). From a small pocket I took out coarse gold grains. This gold is a bright, light yellow with very little oxide, and is not ragged as you might expect from gold that doesn't travel far from its source. Instead, it shows evidence of rounding as if it had travelled some distance. My other test was behind and under a large boulder in the middle of the wash (Fig. 59). This boulder did not sit on bedrock and results were disappointing. The gold from here was very small. All the washes along the mountain flanks contain a hodgepodge of boulders that make prospecting difficult. Unless you can bring in recovery equipment close to where you plan to work, such as the high banks, you may have to contend with sniping bedrock exposures and pack out your dirt as I did. However, if you brought along a metal detector a whole new world opens up to you here.

Narrow quartz veins thread the mountain slopes (Fig. 57). There is much quartz float and numerous prospects and mines. Opportunities are many for discovering hidden gold in quartz and there could be residual placer deposits on the hillsides. It would take several weeks, or months, to cover the area with a metal detector. I metal-detected a small quartz prospect. Although I did not find any gold, I did find buried under the pile of quartz tailings one long steel digging bar, one short steel digging bar, two quartz-picks and one 10 pound sledge, all with a heavy coating of rust. Due to the age of the district, there probably are many such burials of mining artifacts from long ago.

The **Muggins Mountains Placers** are now in the recently established **Muggins Mountains Wilderness Area.** The Muggins Mountains, at one time, were within the U. S. Department of Defense's **Yuma Proving Ground**. Although the map on Fig. 60 indicates the former military boundary, it is now north of the mountains. The Muggins Mountains are directly east of **Dome** on the northern edge of Dome Valley (Fig. 61). These mountains were first prospected in the 1860s. The placer deposits produced considerable gold and silver before 1900, and the total estimated production up to 1942 is 2,748 ounces of gold and about 500 ounces of contained silver. A small number of lode deposits have produced up to 251 ounces of gold, but the main economic mineralization of the district has been the gold placer deposits. Most of the gold placers lie along the southwestern margins of Muggins Peak (Fig. 62), Klothos Temple or Coronation Peak, and Long Mountain (Fig. 63), and in the upper reaches of Vinegaroon Wash. Gold-bearing gravel occurs in washes and canyons draining southwestward from these peaks, mainly in ancient bars on terraces above, and in, the stream channels on a bedrock of conglomerate.

Several months before my planned prospecting entry into the Muggins Mountains

Wilderness Area, I submitted a Plan of Operations to BLM's Yuma District Office. My plan described in detail how and where I would enter the area, what mining methods I would use, what size sample I would take, and the duration of my stay. The acting area manager determined that my activities constituted casual use and did not require a Plan of Operations. I now had the clearance to go on with my prospecting trip with the understanding that, as of November 28, 1990, the Muggins Mountains Wilderness Area had been closed to mining claim location.

If you plan on prospecting in any wilderness area or national forest, I recommend first that you contact the responsible governmental agency in charge and notify it of your plans, as I did in the above example. Those in charge of managing the wilderness areas or the national forests sincerely wish to assist the recreational prospector, and will advise you of the latest rules and regulations governing your proposed mining activities. As long as your activities are not detrimental to the environment, you should not experience any difficulty obtaining a clearance.

Entry into any wilderness area must be on foot. Besides your feet, you may use those of pack animals such as horses and mules. **You may not use any motorized equipment, including helicopters, nor may you use bicycles or hang-gliders, in the wilderness area.** The intent of the law is to keep the wilderness area free of any artificial object detrimental to the ecology and the wilderness experience.

You can access the Muggins Mountains Wilderness Area in a high-clearance, or four-wheel-drive, vehicle in two places. From **Yuma**, travel east on **Interstate 8** to the **Ligurta** exit. Travel eastward through Ligurta to **Dome Valley Road** and turn north. Follow Dome Valley Road (**Avenue 20 E**) to **County 7th Street** and turn east. Go east on County 7th Street past the **Dome Valley Transfer Station** to access the **Muggins Wash** cherry stem. The cherry stem road takes you to the "basin" in which situates the La Mina Victor placer claim. There are other valid claims along Muggins Wash, from its mouth, up to the basin.

I entered the wilderness area by driving up Muggins Wash, *west* of the transfer station, to the signs posted at its boundary (see Fig. A). Going this way requires a four-wheel-drive vehicle. From here I hiked up Muggins Wash to the basin. If you decide to hike up Muggins Wash from the boundary, be careful traversing the steep-walled canyon beyond (Fig. 64). There is no escape from it if there is a flash flood. Scouring of the canyon walls indicates a ferocious and deep flow through this wash in times of heavy rains.

I sniped a bedrock exposure on the main wash and got very small flood gold. Your best bets in this area are the high banks farther up-canyon and the smaller gulches that are not subject to as much flash-flooding, and you will have to pack out your paydirt.

The **Laguna Dam Placers** are at the eastern end of **Laguna Dam**, on the extreme northwestern slopes of the **Laguna Mountains** (Fig. 65). Discovered in the 1860s, this placer contained coarse gold and nuggets in gulches, in potholes in benches as much as 100 feet above the river, and in the Colorado River itself at the eastern foot of the dam. Across the river in California, the potholes were more extensive and mined by individuals well into the 1960s. What is particularly interesting about the California potholes is that, in the 1960s, all a miner had to do was to lay a piece of sheet metal or canvas over his workings and tools to "claim" the spot. Even more surprising is that virtually nobody disturbed his claim, or stole the tools he temporarily left behind. Just try that today!

You can easily reach the Laguna Dam placer area from **U. S. Highway 95** by taking **Avenue 7 E** northward. This is a paved road almost to the placer area. It then becomes a graded gravel road just before you reach the lock as you cross over a small canal. To your right is a wide, flat area at the mouth of a wash. A dirt track leads eastward along the wash. There is evidence of recent mining using heavy equipment to attack the slopes on your left (Fig. 66). I drove farther on to where others were mining a high bank on the right (Figs. 67 & 68). Drywashing piles here indicate the equipment these small miners used to recover the gold from the excavated gravel. From the amount of work done by the miners I concluded that this would be a good place to take a sample. I set up my drywasher on a pile (Fig. 69) and attacked the high bank with my steel digging bar and shovel. Excavating the bank was very easy. The gravel was not cemented except for a thin caliche layer just above bedrock. I processed six hoppers full of gravel and recovered large, very coarse, light yellow gold. This gold was much like that at Las Flores, with very little oxide, and rounded, indicating travel. The largest pieces exceeded one-eighth of an inch in diameter. There was very little in the way of small particles. From my test I concluded that the gold distributes evenly throughout the high bank and that most of it is large. **Sizable nuggets are a distinct possibility here; metal-detect all your screenings!**

THREE PLACERS IN A MINERAL WITHDRAWAL (Figs. 70):

As mentioned earlier, I intended to report on three placers now within the boundaries of the **Kofa National Wildlife Refuge**. The **entire refuge** is a **mineral withdrawal**, which means **no prospecting** and **no locating of mining claims!** Unless you have a valid existing right before the establishment of the mineral withdrawal, such as a claim in good standing, or patented property, you may not do any mining in the withdrawn area. Unfortunately, I had neither a claim nor patented property in the game refuge, so prospecting was out of the question.

The **Kofa Mining District** (Figs. 71 & 72) lies in the southern part of the **Kofa Mountains** of central Yuma County. The name for the mountains is a contraction of "King of Arizona," the most prominent and famous lode gold mine in the district. An earlier, and perhaps more vulgar name for the mountains, still used, is the S. H. Mountains.

Charles E. Eichelberger discovered the King of Arizona vein during the winter of 1896. He, with H. B. Gleason and Epes Randolph, organized the King of Arizona Mining Company. In 1906, Felix Mayhew discovered the North Star vein 1-3/4 miles north of the King of Arizona mine. The combined total value of gold produced by these two mines exceeds 40 million dollars at today's bullion prices! Interestingly, neither of the veins contributed to the placer gold found in the washes near the lode mines. The belief is that numerous disintegrating gold-bearing veins in the metamorphic rocks led to the placers, now localized in a small area defined by North Star Wash, lying between the King of Arizona and North Star mines.

To reach the Kofa Mine area, drive southeastward from **U. S. Highway 95** on **King Road**. This signed road connects with the highway a few hundred feet north of the dilapidated gas station at **Stone Cabin**. From here, it is 19.4 miles on a good dirt road to the intersection at the King of Arizona Mine (Fig. 71). When you reach the refuge boundary, there is an information center from which you can obtain a map of the refuge and the latest regulations (See Appendix). Although the map indicates King Road suitable only for high-clearance vehicles, I believe that, with care, you can negotiate it in a passenger car. However, most designated roads in the refuge require a four-wheel-drive vehicle. This is a very hostile environment and there are *no* services available within miles if your vehicle breaks down. Always take along plenty of water, particularly if you go into the area during the summer months. Because of the intense summer heat, I do not recommend visiting this refuge from May through September.

At the time of my visit (September 20, 1993) there was active placer mining going on at the King of Arizona Mine locale. The King of Arizona and North Star mines are on patented land (Figs. 73 & 74). There are many open shafts, hundreds of feet deep, throughout. I do not advise wandering about, and I, especially, do not recommend letting children run loose in this area.

The **Castle Dome Mining District** (Figs. 75, 76 & 77), situated in the southern portion of the **Castle Dome Mountains**, was primarily a lead-silver lode mining district that included values in gold, copper, zinc, molybdenum, vanadium, fluorite and barite. This is one of the oldest mining districts in Arizona. The earliest American prospectors, who discovered the lode deposits in 1862, found evidence of still earlier mining by others. The identity of these earlier miners is unknown. Although this district is not notable for placers, a small amount of gold generally occurs with the primary ores (see **Patagonia Mountains** and **Plomosa (North) Mining District**). If this gold fortuitously accumulates in the gravel, important placer deposits could exist. Right now, there is no legal way of determining this if you do not have a valid prior existing right.

To reach the Castle Dome Mining District, drive northeastward on the paved road that intersects **U. S. Highway 95** just south of the **Aerostat** facility. A sign at the highway clearly marks this road. Beyond the Aerostat facility the road turns into a good gravel road passable in a passenger car. When you reach the area around Castle Dome Peak you can see numerous tailings piles on the low slopes to the Southeast. What you don't see are the dozens of open shafts in the flats. This is another very dangerous area not suitable for children. Be extremely careful and take along plenty of water if you do any hiking here!

The **Castle Dome Gold Placers** are near the **Big Eye Mine** on the eastern slope of the **Castle Dome Mountains**, principally in the small washes draining down from the mine into **Big Eye Wash** (Fig. 78). The Big Eye lode mine, located near the head of Big Eye Wash, had ore containing spotty values of gold and silver along with minor amounts of copper, lead and zinc. The disintegration of numerous gold-bearing veins nearby was probably responsible for the accumulation of placer gold in the gulches. Prospectors discovered the placer deposits in 1884 and worked the gravel sporadically with drywashers before 1900. After a lull of many years, mining activity renewed in the 1930s during the Great Depression, then declined over the next several decades. The total estimated gold production from the placers up to 1900 is about 7,000 ounces.

To reach the Castle Dome Gold Placers and the Big Eye Mine, take the four-wheel-drive dirt road branching southeastward from the Castle Dome road. A sign clearly marks this road. I do not recommend taking a passenger car on this road although it is passable to one for the first couple of miles.

I had tremendous enjoyment testing and sampling the placer fields outlined in this chapter, particularly the more remote fields, and I got good gold. I hope you have as much fun and success as I did when you go out to these placers. Good luck and good hunting!

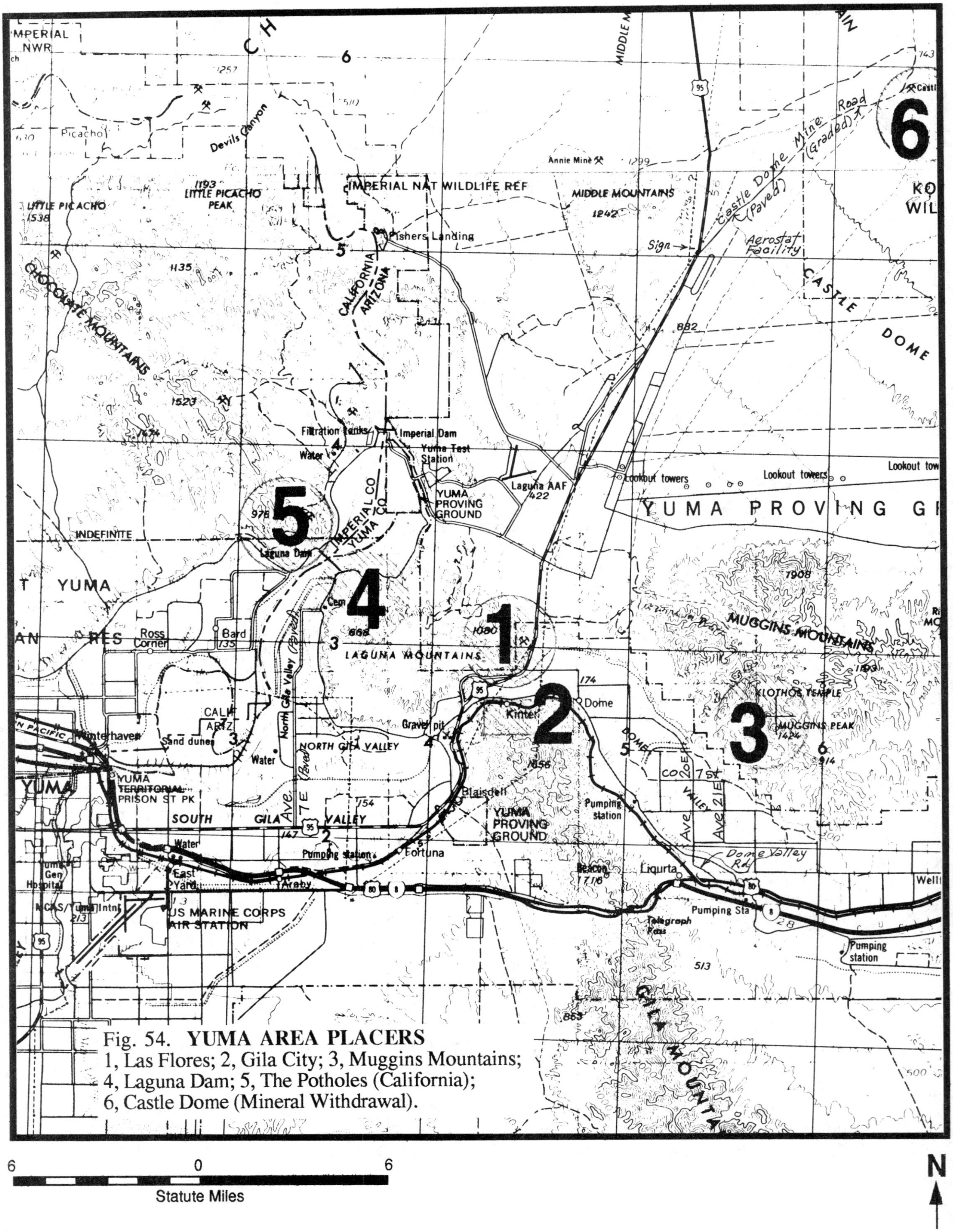

Fig. 54. **YUMA AREA PLACERS**
1, Las Flores; 2, Gila City; 3, Muggins Mountains; 4, Laguna Dam; 5, The Potholes (California); 6, Castle Dome (Mineral Withdrawal).

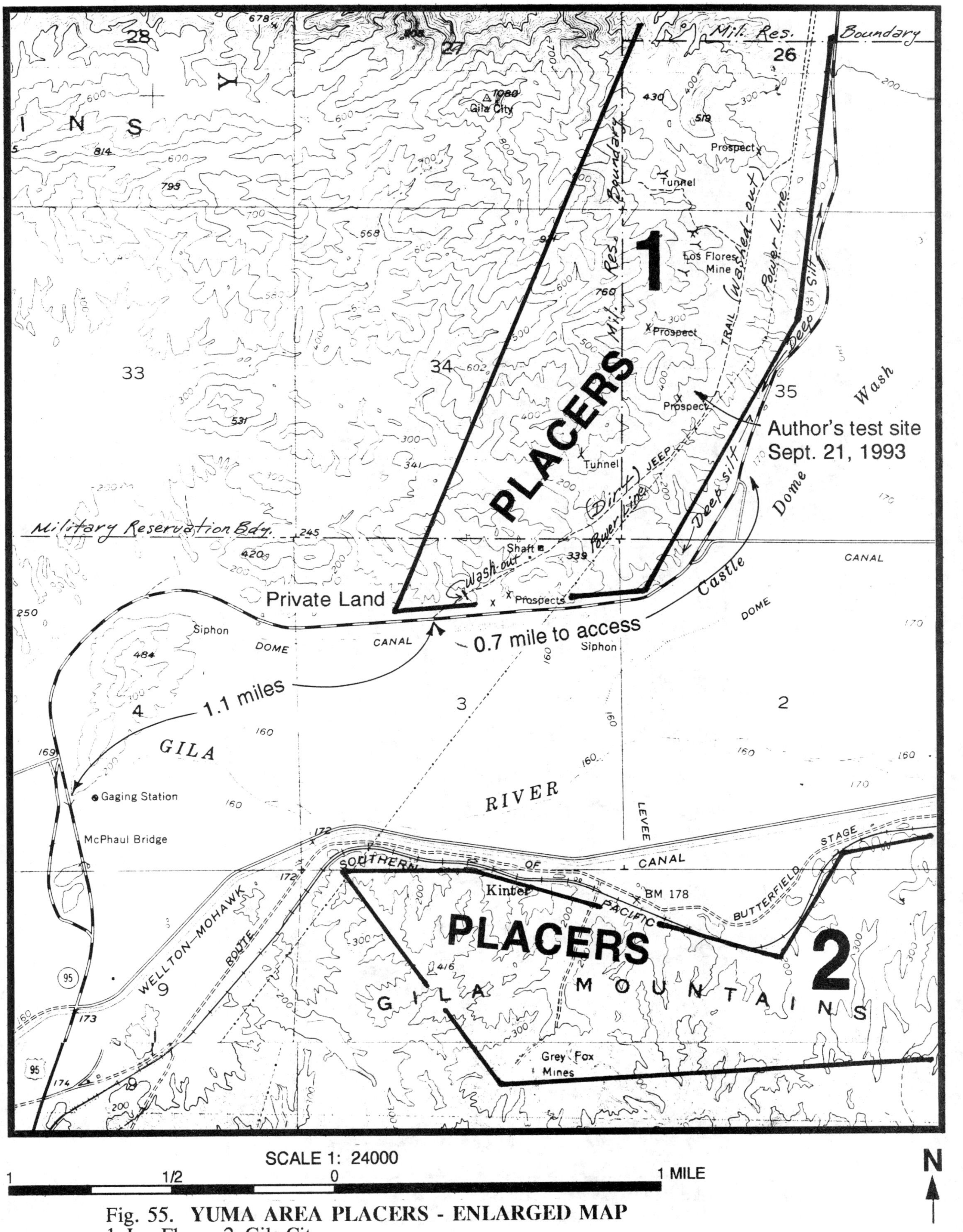

Fig. 55. **YUMA AREA PLACERS - ENLARGED MAP**
1, Las Flores; 2, Gila City.

Figs. 56 & 57. **LAS FLORES PLACERS**
56, Southeastern flanks of the Laguna Mountains; 57, Quartz vein in schist, and float (arrows).

Fig. 58. **LAS FLORES PLACERS.** Author's equipment set up for sniping a pocket behind a bedrock outcrop in a dry wash, September 21, 1993. The point of the shovel indicates where I began the excavation. From this small pocket I recovered pin-head-size, coarse gold grains of a light yellow color. Although the source of the gold is the many veins and stringers on the mountain in the background, the gold is not ragged, but rounded instead.

Fig. 59. **LAS FLORES PLACERS.** Author's equipment set up for sniping behind a large boulder in the same wash. This boulder did not sit on bedrock and the gold I recovered here was very small. Note the piles of screened tailings, left by previous miners, just behind and above the handle of the shovel in the photo. These miners had mined the benches and I estimate the age of the piles at 60 years old, or in the time of the Great Depression of the 1930s. A dirt road, washed out in many places, cuts across the upper half of the photo.

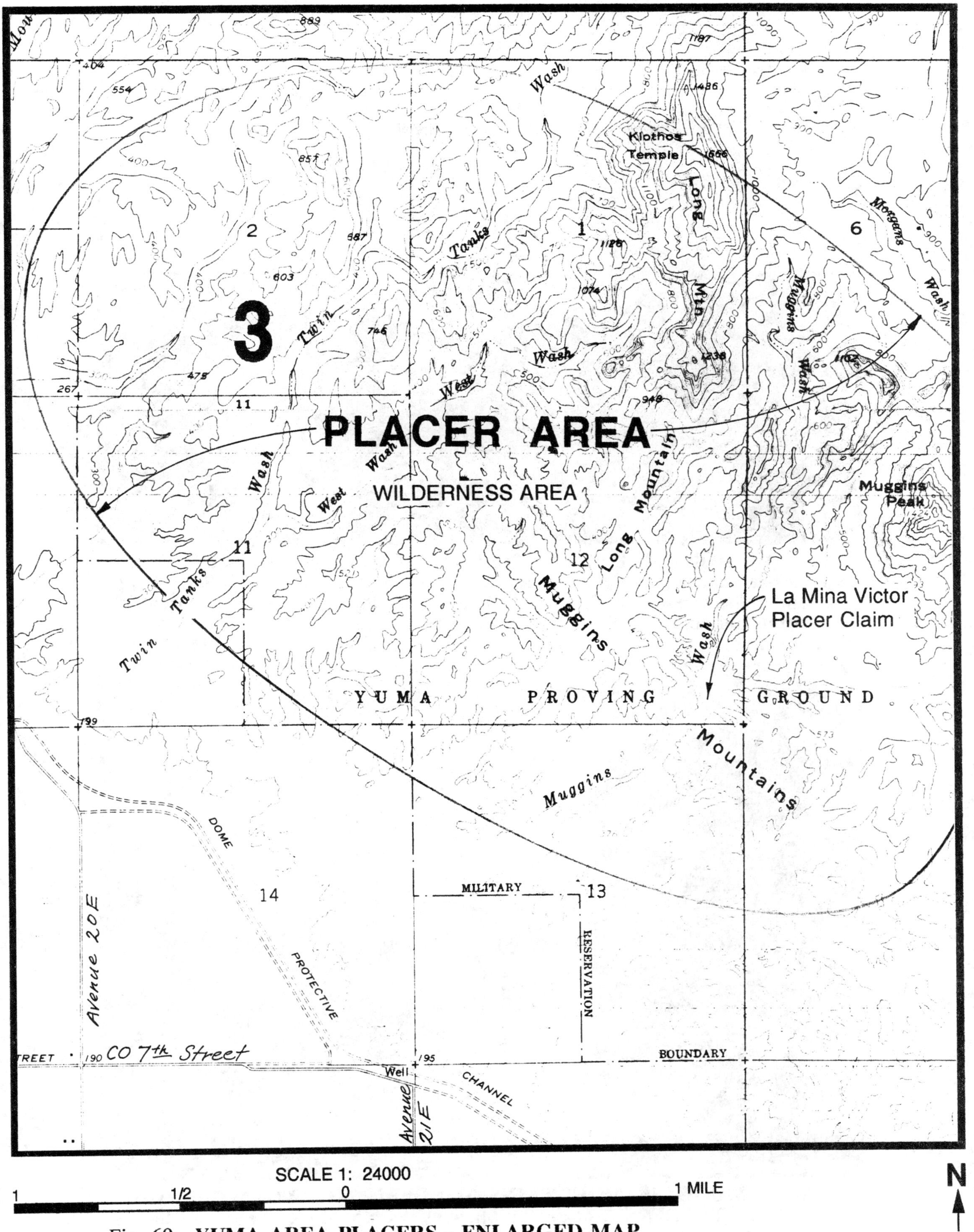

Fig. 60. **YUMA AREA PLACERS - ENLARGED MAP**
3, Muggins Mountains Wilderness Area.

Fig. 61. **MUGGINS MOUNTAINS PLACERS.** The Muggins Mountains as seen from the intersection of Dome Valley Road (Avenue 20 E) and County 7th Street, with crop land in the foreground, September 22, 1993. To the left-center of the photo is Long Mountain, and at right-center is Muggins Peak. This mountain range is a wilderness area.

Fig. 62. **MUGGINS MOUNTAINS WILDERNESS AREA - MUGGINS MOUNTAINS PLACERS.** Looking up Muggins Wash to Muggins Peak, September 22, 1993. Farther up the wash I tested a bedrock exposure and got finely-divided flood gold.

Fig. 63. **MUGGINS MOUNTAINS WILDERNESS AREA - MUGGINS MOUNTAINS PLACERS.** Looking north to Long Mountain, September 22, 1993. The washes draining off this side of the mountain carry small particles of placer gold.

Fig. 64. **MUGGINS MOUNTAINS WILDERNESS AREA - MUGGINS MOUNTAINS PLACERS.** Looking up Muggins Wash to the entry slot to a steep-walled canyon from which there is no escape if it rains. Note the scouring of the side banks from flash floods.

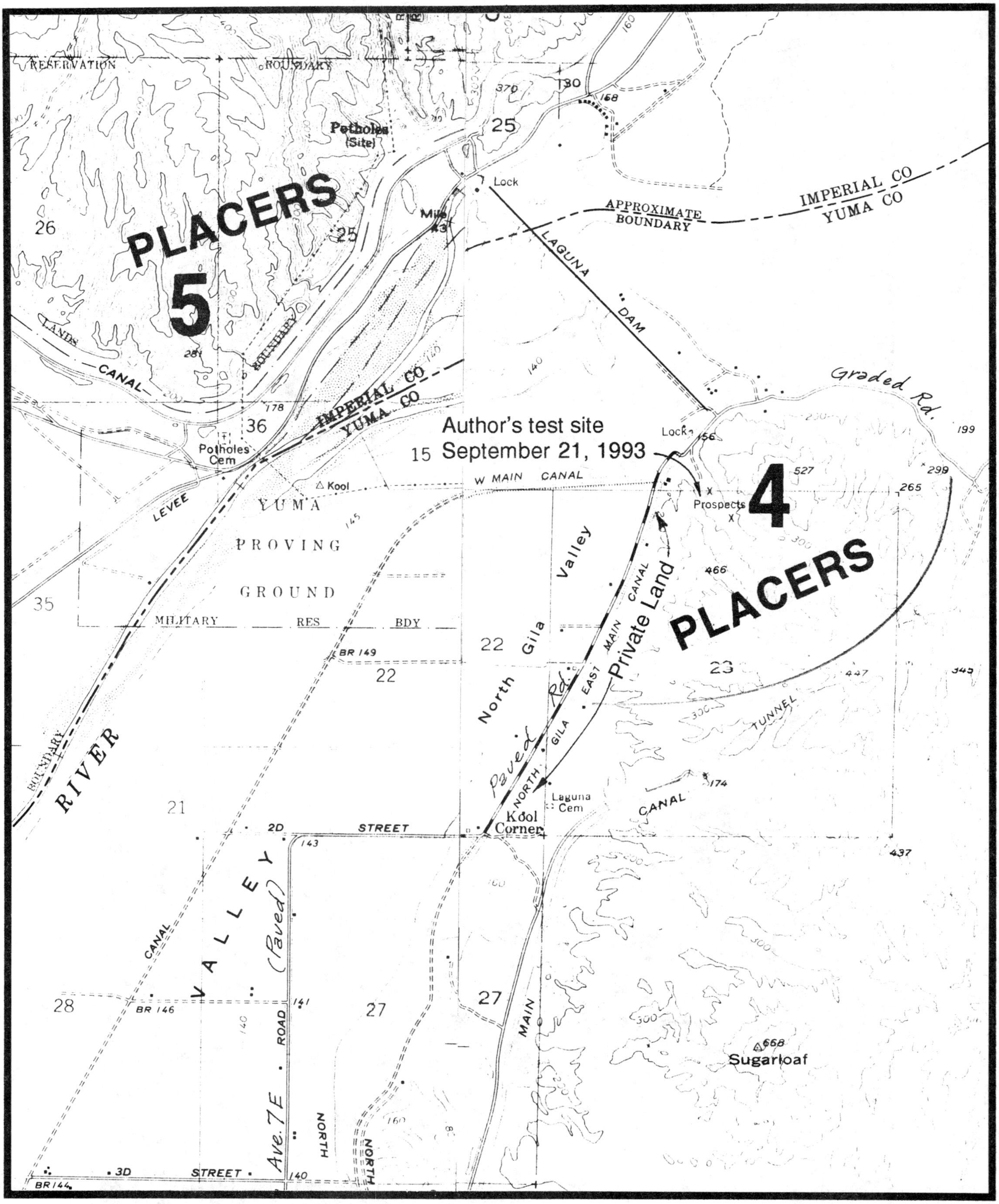

SCALE 1: 24000

1 1/2 0 1 MILE

N

Fig. 65. **YUMA AREA PLACERS - ENLARGED MAP**
4, Laguna Dam; 5, The Potholes (California).

Fig. 66. **LAGUNA DAM PLACERS.** The scars of large-scale mining of the lower flanks of the mountains, just east of the lock, using heavy equipment, September 21, 1993. Although there was no activity at the time of my visit, it appears that mining does take place here on and off.

Fig. 67. **LAGUNA DAM PLACERS.** A high bank showing recent excavation by recreational gold miners using drywashers for gold recovery. It appears that continuous mining over the years has cut back the high bank a distance of several yards.

Fig. 68. **LAGUNA DAM PLACERS.** A close-up view of one of the excavations in the high bank, September 21, 1993. Note the drywasher pile in the foreground. I sampled this high bank by enlarging the excavation and processing the gravel through my drywasher. From here, I took out large, coarse, light yellow gold nuggets.

Fig. 69. **LAGUNA DAM PLACERS.** The author's set-up at the high bank with his drywasher at the excavation, September 21, 1993. This high bank appears to be a debris flow, or landslide, when seen from across the wash.

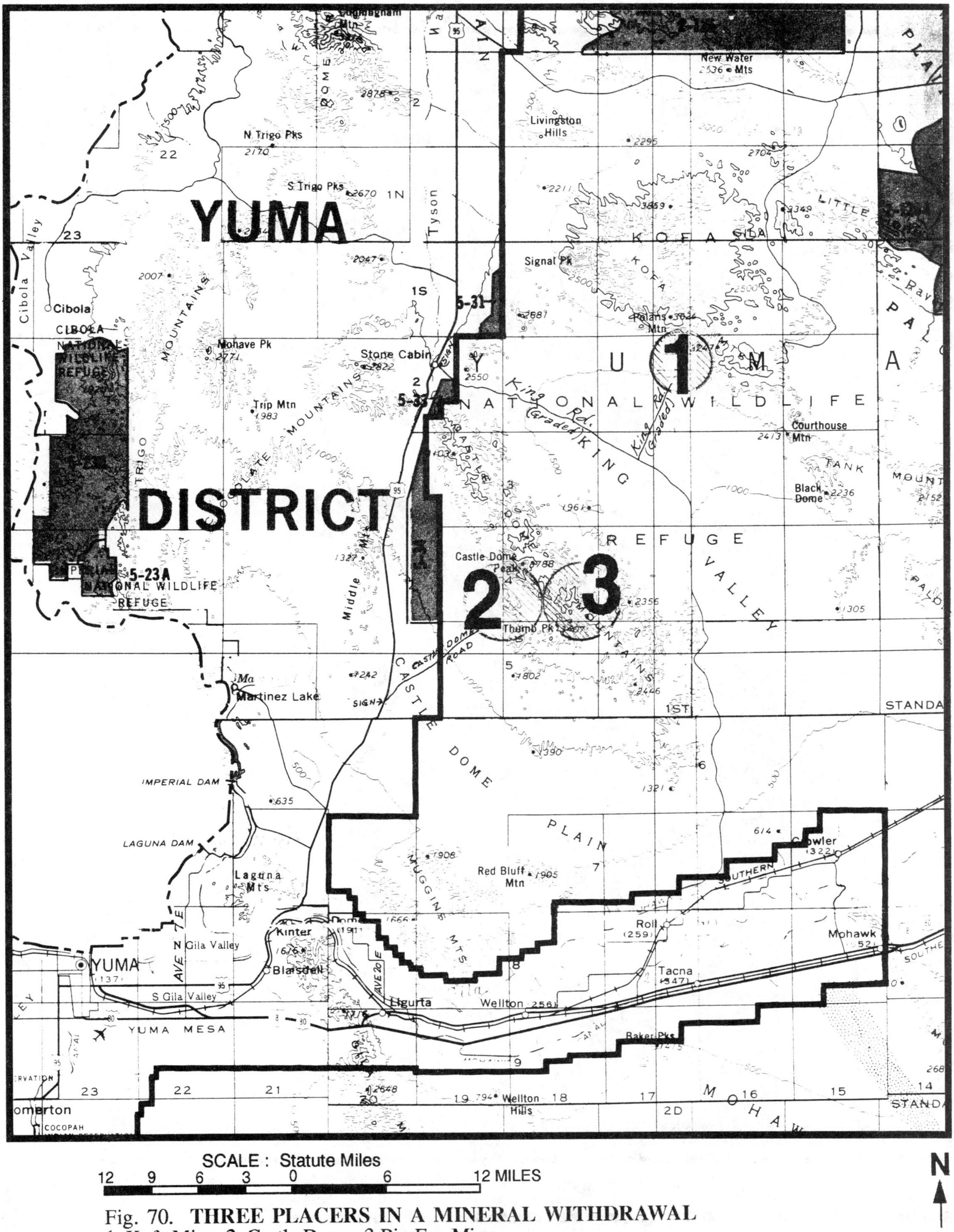

Fig. 70. **THREE PLACERS IN A MINERAL WITHDRAWAL**
1, Kofa Mine; 2, Castle Dome; 3 Big Eye Mine.

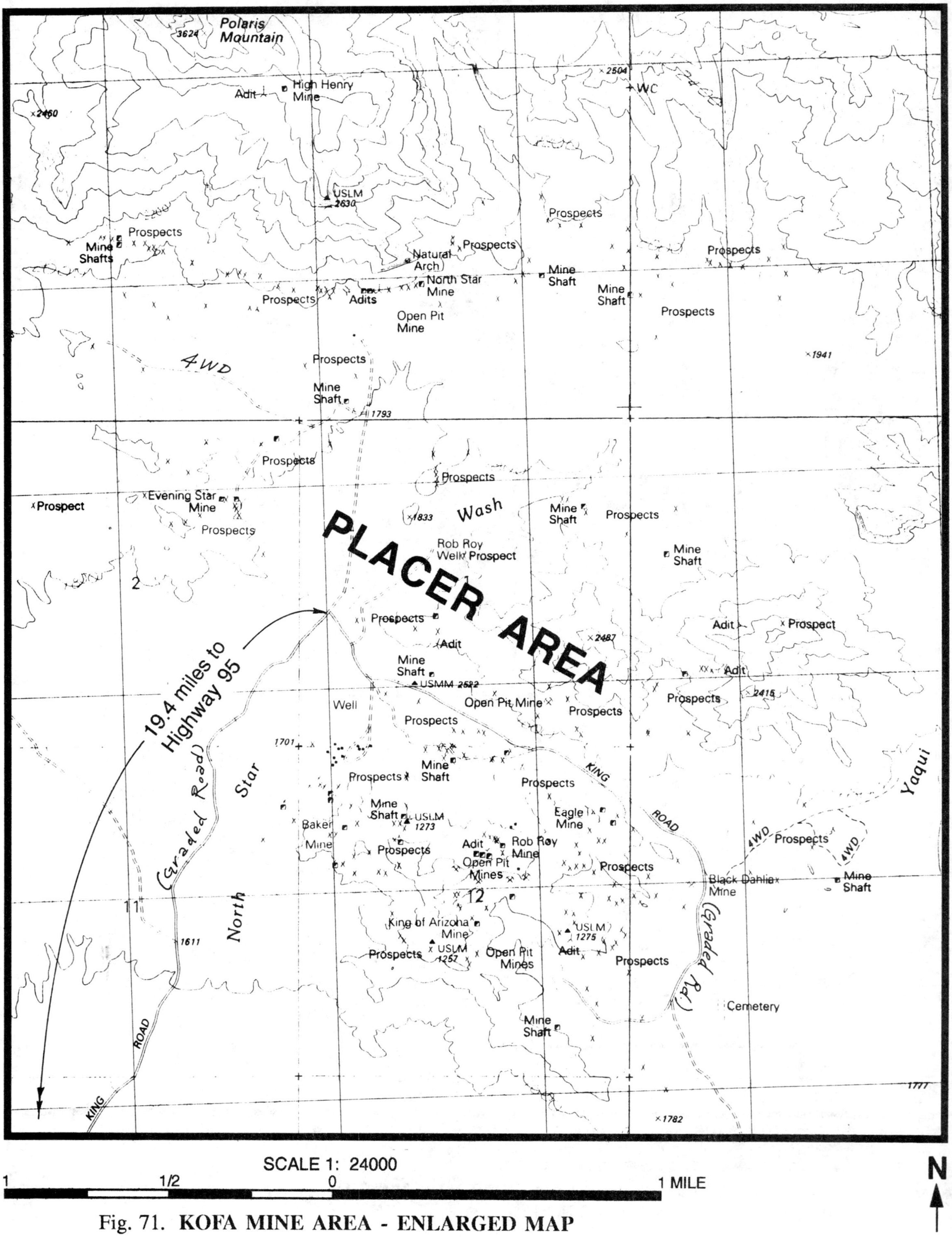

Fig. 71. KOFA MINE AREA - ENLARGED MAP

Fig. 72. **KOFA NATIONAL WILDLIFE REFUGE.** In the distance are the Kofa Mountains as seen from King Road, looking northeast, September 20, 1993. The Kofa Mining District is at the base of the mountains just to the right of the center of the photograph.

Fig. 73. **KOFA MINING DISTRICT.** This is the fabulous King of Arizona Mine, one of the biggest producers of lode gold in Arizona, now occupied by a few rugged souls. All the land here is private property. There are numerous, deep, open mine shafts in the vicinity.

Fig. 74. **KOFA MINING DISTRICT.** Polaris Mountain, Kofa Mountains, Kofa National Wildlife Refuge, September 20, 1993. At the base of the mountain, just a little to the right of the center of the photo, is the tailings pile of the North Star Mine, a major past-producer of lode gold. Mining activity, here, still continues to this day.

Fig. 75. **CASTLE DOME MINING DISTRICT.** In the distance are the Castle Dome Mountains as seen from Highway 95, looking to the Northeast, with Castle Dome Peak at upper left.

Fig. 76. **CASTLE DOME MINING DISTRICT - KOFA NATIONAL WILDLIFE REFUGE.** Castle Dome Peak is the upper left of the photo, September 21, 1993. There are many open mine shafts from here to the base of the mountains. Placer gold probably occurs in all the washes draining this area, but prospecting is forbidden unless you have a valid prior existing right.

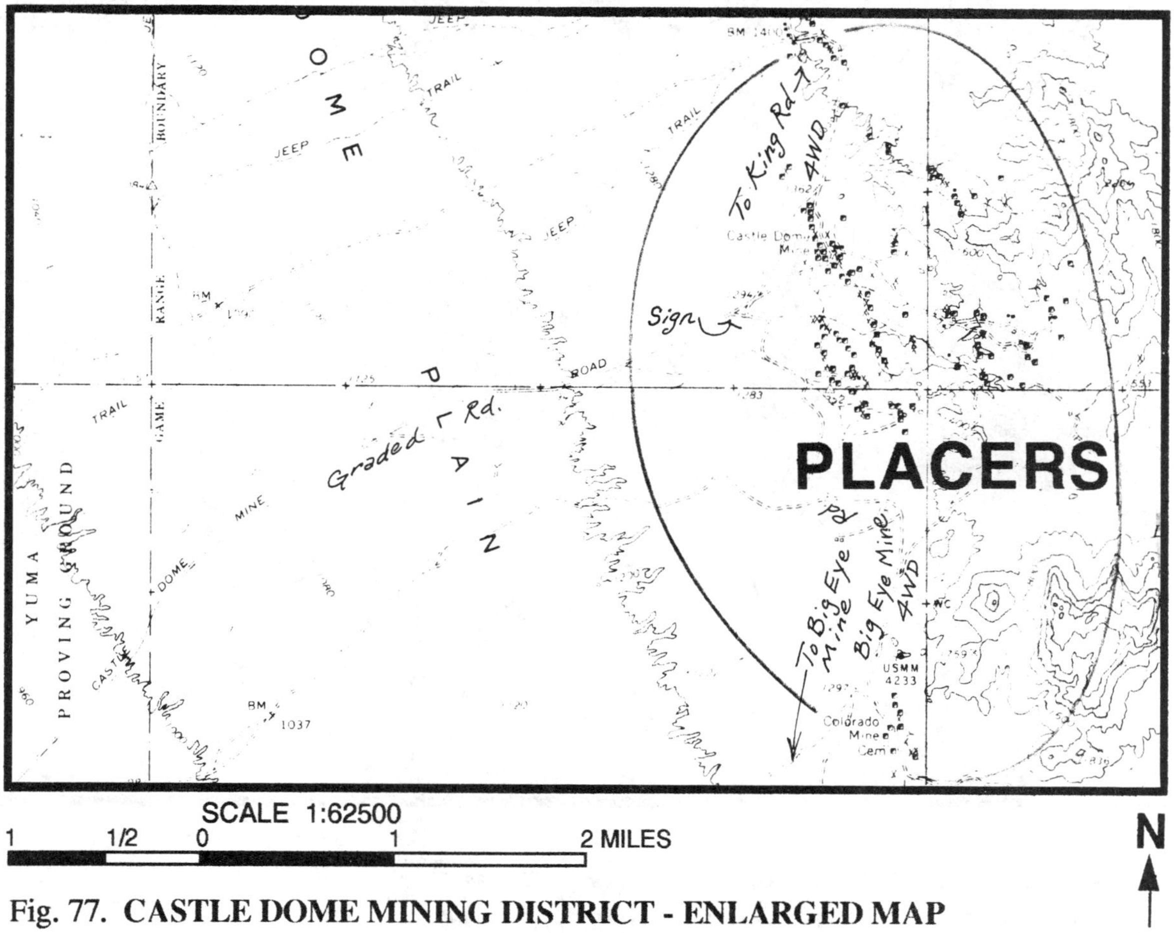

Fig. 77. **CASTLE DOME MINING DISTRICT - ENLARGED MAP**

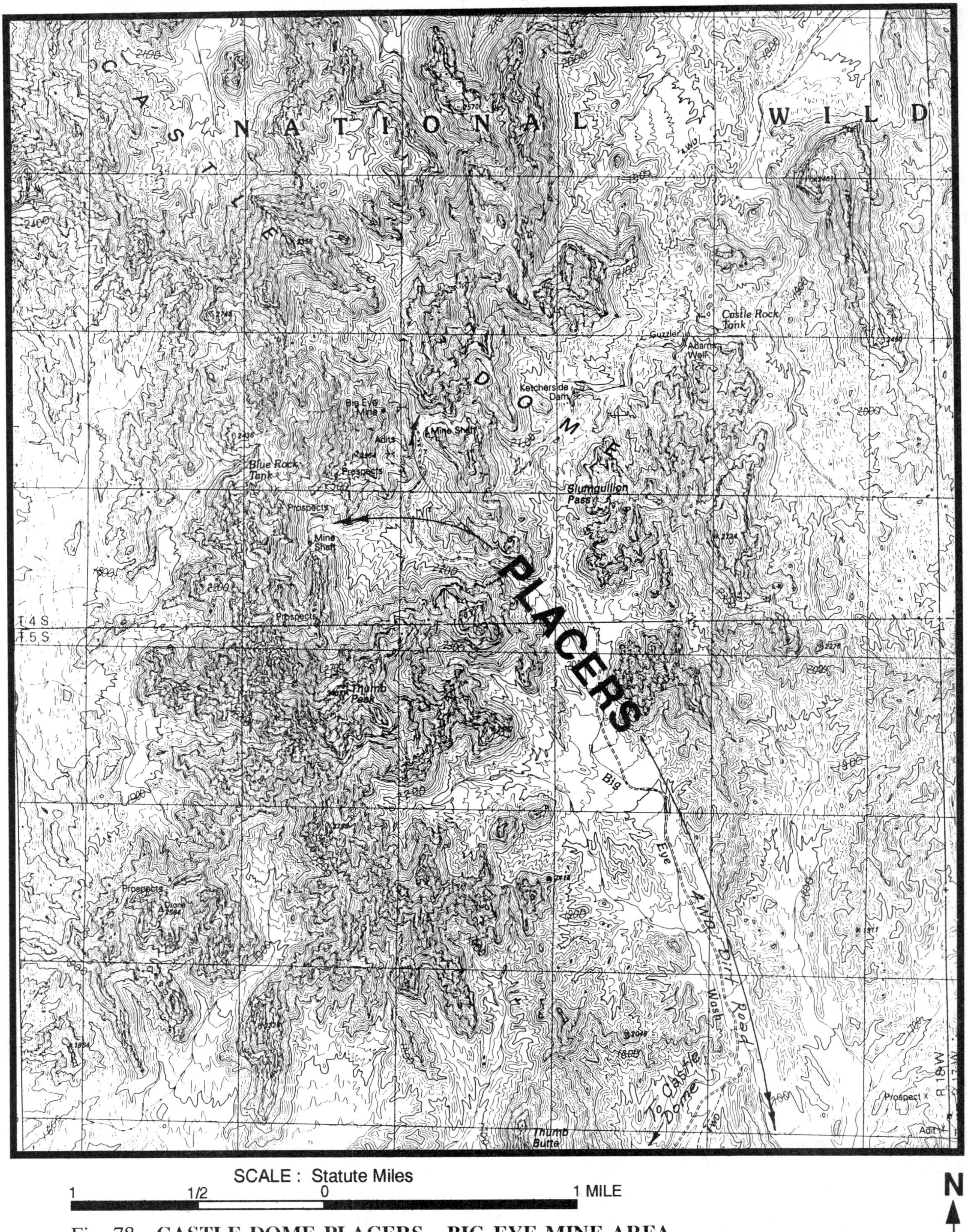

Fig. 78. **CASTLE DOME PLACERS - BIG EYE MINE AREA**

CHAPTER II

SNIPING ON A SHOESTRING

In my book ***Arizona's Golden Secret***, copyright 1992, I discussed briefly a method of mining known by most experienced prospectors as "sniping." Sometimes it goes under the name of "crevicing." However, by whatever name you call it, it is a way you can get gold with only a minimum investment in equipment and time. The object here is to apply careful observation of field conditions, backed up by knowledge of what gold does when moved by flowing water, to get gold overlooked or bypassed by others.

We know that, in flowing water, gold sinks downward, inexorably, until stopped by something, usually bedrock. It may sink into a crack or crevice and stay there for a very, very long time until moved forcefully by flowing water or pried out by the savvy gold hunter. It tends to hang up behind obstructions in the eddy current where the flow slackens.

The riffles in a sluice box take advantage of these characteristics of gold to capture it. What you want to do, then, is to look for those natural conditions analogous to the sluice box's riffles. You want to concentrate your efforts toward prospecting for cracks and crevices, fissures and ridges, and dips and pockets in the bedrock underlying the gravel bed of small tributaries, streams, creeks and washes. You should look for obstructions to the flow that create eddy currents on the downstream side, or any other natural feature of the watercourse causing a slackening of the flow. All are conditions that temporarily halt gold's incessant march downstream and what you, as a gold hunter, should be looking for.

In some Arizona gold-bearing areas, unusually heavy and prolonged seasonal rains will bring in a fresh crop of gold from the eroding hillsides and deposit newly moved gold into the same cracks, crevices, fissures and pockets. In this case, you can take advantage of an occurrence that may come once in several years. However, you will have to be Johnny-on-the-spot to get that gold before others. This is especially so regarding the lower, more accessible, heavily worked gulches. In these popular spots you should avoid going over ground that miners work repeatedly as the rewards are few. You can usually recognize these places by the remnant tailings left by previous gold hunters. It is only after several years have elapsed with no mining activity in these areas would it be worth your while to try your luck.

To maximize your efforts as a sniper, you should look for untouched, or virtually untouched situations overlooked or bypassed by other miners. These spots, such as the upper tributaries of gold-bearing creeks and washes, were relatively difficult for earlier miners to get to, and work, with their mining equipment. Many times, they passed them by in favor of lower, richer creek and stream gravel easily processed by their equipment. Most earlier placer miners used rockers and sluices, devices dependent on availability of water and sufficient gravel for processing, conditions usually lacking on bedrock situations of small tributaries and intermittent creeks. Their need was to earn a daily living, unlike today's recreational miner, so they concentrated their efforts on the richest portions of gold-bearing streams and washes and ignored ground they considered too low-pay or too much effort to work. At today's gold prices, this ground is worthy of our attention.

Many of today's miners want easy access to the gold-bearing creeks and washes, places where they can drive to, park alongside and drop in their gold recovery equipment. Also, their equipment may require a reliable water supply to process the gravel, therefore limiting them to the larger creeks and streams. They bypass many of the good gold situations thus leaving these for the sniper. However, there are tradeoffs so that sniping is no more "better" than other mining methods. In this case, you are dealing with highly selective mining, applying to the fullest all your experience and knowledge to recover the most gold from the *least* amount of material. Many other mining methods rely on processing a *large* volume of material to recover the most gold. All methods have their place when it comes to getting gold.

Before you start out to prospect for the best spots to do some sniping you should have on hand the following equipment: small folding shovel; prospector's pick; miner's pick; pry bar; chisel; small sledge; trowel; spoon; screwdriver; screen (1/4" mesh; but, watch out for larger nuggets too big to pass through the screen); whisk broom; knapsack or plastic bucket and your gold pan. Not all situations demand that you have on hand all the tools listed. You probably could get by with just a shovel; trowel; spoon; screwdriver; whisk broom; gold pan; and a container for your paydirt. It is likely that you will have to hike to some of the better gold-producing spots and you will have to carry in your food and drinking water. A backpack frame and backpack to carry your equipment is necessary as well as a good physical condition. Barring any serious physical disability, you probably will come out in better physical condition doing hiking and sniping than before. One last item deserving attention by the serious sniper is a gasoline-powered vacuum cleaner for cleaning out crevices. I discuss this at length later in this chapter.

Before you embark on a sniping expedition you should select your gold-bearing area carefully. The best place is a known gold placer area characterized by much exposed, easily accessible (little or no overburden), fractured bedrock with many cracks, fissures and pockets. One of the best is the Greaterville placers. Another is the Las Guijas placers. Each area has bedrock very close to the surface in the upper tributaries with very little overburden. Each has bedrock distinguished by many cracks and fissures running perpendicular to the flow of the streamcourses. Additionally, each has enough bedrock exposed on the sides of the banks to enable you to read the bedrock and form some ideas about what it does under the overlying gravel. These two placer fields offer some of the best terrain for the novice sniper to learn his craft and reap a reward for his efforts. A third area is the Lynx Creek Recreational Panning Area near Prescott set aside by the U. S. Forest Service specifically for the recreational prospector. The Forest Service permits only shovels and pans, that is, no motorized equipment allowed, but I know of no specific prohibition against using trowels, spoons, screwdrivers or whisk brooms for cleaning out bedrock crevices. My best efforts here resulted from cleaning out exposed bedrock cracks and crevices on the high banks above the creek. The cracks here are deep and you should have a pry bar to open them.

Although I have given three examples of good sniping terrain, they are not the only areas where you can apply this mining method. The principles of sniping are applicable to any gold placer area. You need to know these so that you can go into any gold-bearing placer area and be able to apply them successfully. They should be a part of your arsenal of mining knowledge that allows for maximum flexibility and options on your part.

Let's start out in an area noted for good sniping terrain, Greaterville, in the Santa Rita Mountains south of Tucson. I have done much sniping and crevicing of the upper tributaries here and can personally testify to the validity of these techniques. Most of the miners I have seen here restrict their efforts to the major streamcourses that are easily accessible by vehicle. Untouched spots are hard to find as a result of miners working the ground repeatedly. Sometimes, heavy rains bring in a new crop of gold to these lower streamcourses, or rearrange existing gold. To get this new gold, you have to know, in advance, the best spots, and be there before others. What we want are spots that have gold accumulating over many, many years and are untouched, or relatively untouched, since the early days. From my observations, very few miners work the upper tributaries.

The major mineralized zone in the Greaterville placers is Granite Mountain, an igneous intrusive, rising about a mile west of Greaterville. This low mountain has many lode mines that are visible from the dirt roads traversing the area, and most of the major gold-bearing tributaries and gulches head into it. The most notable gulches are Empire, Chispa, Colorado, Los Pozos, Ophir (Hughes), Louisiana, Sucker, Kentucky and Boston. Other gulches are Enzenberg, Nigger, St. Louis, Harshaw, Graham and Fish. I have tested bedrock pockets on Ophir (Hughes) Gulch and found good coarse gold, probably of recent origin as gold hunters heavily work this area and untouched spots in the stream bed are rare. These lower, more accessible gulches are better for high banking, described in detail in Chapter III.

Dirt roads encircling Granite Mountain reach most of the upper tributaries. Many of the roads on the west side of the mountain are negotiable only by four-wheel-drive vehicles or pickup trucks with high centers. One road that is passable in a passenger vehicle (FH 163) goes south from Melendrez Pass Road (FH 165) just west of the Morning Star Mine on the east side of the mountain. This road allows you reach the upper tributaries of most of the major gold-bearing gulches with only short hikes from your parked vehicle. One of my favorite gulches is Sucker, which has three small tributaries heading into Granite Mountain just below the dirt road. Although narrow, these tributaries have much exposed bedrock and very little overburden. They are winding, enough so to provide many opportunities on the inside bends, but the real attraction is the bedrock that forms many cracks, fissures and pockets.

The uppermost parts of these tributaries are steep and narrow with limited potential but the lower parts flatten and widen a bit to allow the gold to settle. What I look for is not the exposed, scoured bedrock but bedrock that has a covering of gravel. This indicates a slowing of water flow and potential deposition of gold. A good indicator is where the bedrock exposes on the side banks, showing its character, and traceable under the gravel. When I find such a spot I lay down my equipment and plan out how I will begin mining. The first thing I do is remove the barren overburden from a small area to expose the paydirt. In these upper tributaries the paydirt is very thin, only a couple of inches at the most. This paydirt usually contains gold that is small but there is always the possibility of finding a large, stray nugget in it. To mine it on a bigger scale takes us out of the realm of sniping. Since we are going to have to carry out our paydirt we have to be selective and concentrate our efforts on the cracks, crevices, etc.

Next, I want to expose the underlying bedrock to determine its character. Is it smooth or cracked? Does it present obstacles in the gold's path? Are there gold traps? If, I like what I see, I begin mining. I'll shovel off a large section of overburden to expose the bedrock cracks. Sometimes I come across a pocket that contains compacted paydirt. The gold here is larger and more concentrated so I remove and save all the paydirt from the pocket. By screening it carefully, I am able to reduce the amount of dirt I will have to carry out. I usually use a one-quarter inch mesh but I check out, very carefully, all material not passing through the screen for larger nuggets. Because the potential is great for finding larger nuggets, I take along my metal detector to check out all screenings. I also use the metal detector to go over the thin covering of paydirt for nugget detection before I shovel it off the bedrock.

When I find a number of cracks and crevices, I want to remove all the dirt found in them. Some cracks are very narrow and I open these with a chisel and sledge. A crow bar makes a good pry bar because of the fulcrum that allows for a great levering action. A long thin screwdriver helps to pick and remove the dirt, and the gold, wedged in the crack. You must remove all the dirt in the crack down to the very bottom otherwise you might leave behind some very good nuggets.

Some cracks are wide and shallow. A long-handled metal spoon may be all you'll need to clean out the paydirt. Use your whisk broom to sweep the bottom of the crack clean. Save all the dirt. You usually cannot see the nuggets as dirt covers them. You will have to pan out the dirt to reveal the nuggets. Don't ignore any crevice or crack that shows it contains dirt. Clean out every one! Remember that once gold drops into a crack or crevice, fissure, etc., it will remain there for a very, very long time, or until you or the forces of nature remove it. I would rather it be you than the forces of nature.

Another good situation occurs quite frequently in these upper tributaries. Here is where an outcrop of bedrock sticks out into the tributary from one side or the other. This outcrop creates an eddy current behind it where the gold drops out. You should dig behind (downstream from) this outcrop. Many times there is a small, deep pocket here where the outcrop plunges downward into the gravel. Take out and save all the compacted gravel, down to the very bottom of the pocket, after removing the barren overburden. Expose the bedrock downstream of the outcrop on the outcrop side to reveal any cracks. These cracks usually contain good gold.

Occasionally you will come across a lone boulder sitting in the middle of the tributary. If it's large, then it may have sat there for a long time, accumulating gold behind and under it for many years. It will pay you to take out all the dirt behind and under this boulder down to the bedrock. You can use a pry bar to move the boulder out of the way or a come-along with a length of heavy steel chain wrapped around the boulder. It will be easier for you to move the boulder downgrade and you may have to remove some overburden behind it to provide a pathway. Smaller boulders probably are recent arrivals with little gold behind and under them. If you can move these using both of your arms, ignore them.

Sometimes you will see tree roots exposed on the bedrock. These are good indicators of bedrock cracks in the immediate vicinity. The roots of trees seek out the moisture that tends to remain in the cracks and they will grow in and enlarge the cracks. By carefully tracing the roots along the surface you can discover the cracks. Open up the cracks and remove the tree roots growing in them. Shake out the roots into your container as the fine roots are excellent gold traps. Tree roots are one of the best gold situations you can find because they have remained in place for many years, possibly decades. More than likely, others ignored them because of the amount of work it takes to get the gold out. Human nature being what it is, even the early day miners wanted easy pickings, more gold for less work.

The gold in these upper tributaries is very coarse, ragged and covered with much reddish-brown to black iron oxide indicating a close source. It is also large, nuggets exceeding one-quarter inch not uncommon. This is also true for the tributaries in the Las Guijas Mountains but here the oxide cover is reddish to salmon-colored. The gold from the Lynx Creek area is a very bright, brassy yellow and that from the upper bedrock cracks has some quartz attached. The gold from bedrock cracks in the Huachuca Mountains is coarse, ragged and "lumpy," deep in color indicating a high copper content, and often has grayish quartz attached. Each gold placer area has gold that is distinct from that of other areas. Coarse and ragged gold with attached oxide and quartz indicates that the gold has not travelled far from its place of origin. All these characteristics give you some idea of the host rock in which the gold originally formed and is particularly useful if you were searching for the source, or lode.

I mentioned bedrock exposures above the creeks as worthy of your mining efforts. When you look at these exposures, imagine a period back in time, perhaps millions of years ago, when the creek level was above this bedrock. In other words, in your mind's eye, fill up the gulch with dirt to the level of the bedrock and imagine that the creek was flowing not in its present location but on this bedrock. Over the millions of years that the creek flowed, floods and surges brought in gold from eroding veins and stringers, over this bedrock, and into its cracks. The time scale on which this occurred was immense and the amount of gold accumulating hard to imagine. When you see such an exposure of bedrock you may be into some good gold. The hidden bedrock, that covered by hillside detritus, holds the most promise and the most work. By removing the layer of dirt and gravel covering this bedrock you can expose the cracks perhaps never before mined.

There is much gold in the upper tributaries but the lack of water virtually excludes equipment such as suction dredges, which are dependent on water being available in the tributary itself, and high bankers and mini sluices. To use the last two, you either have to carry in your water and recycle it or carry out your paydirt to your equipment where you have water available. Neither choice is realistic due to the weight of water or dirt you must carry. Sniping appears to be the only way to get the gold from these upper tributaries. However, you can do sniping one other way by using a metal detector designed to detect gold. Since metal detectors are expensive items, it wouldn't be exactly sniping on a shoestring.

Perhaps you have the financial resources to invest in a metal detector to prospect for gold. In that case, buy the best detector you can with the money you have. Your local metal detector dealer will be a great help to you in determining the detector most suited to your needs and budget. Try to get a detector with hot rock identification and the ability to detect very small nuggets. A visual identification meter, although helpful, is not necessary. Use your detector with headphones to screen out extraneous noise, enhance weak signals and prolong battery life. Do not use any discrimination. A small nugget-shooting coil works best when investigating cracks with a detector. With a metal detector, you will be able to check out the shallower cracks for the presence of gold nuggets. It will allow you to check quickly cracks in scoured bedrock without first having to spend time cleaning these out. I usually ignore scoured bedrock as this indicates a swift flow that doesn't allow the gold to settle out, but there is always the possibility of a vagrant nugget being trapped in one of these cracks. Also, the detector really shines when you want to check out the paydirt on the bedrock, and the gravel too large to pass through the screen you use.

Remember this: Gold nuggets give soft responses; **if in doubt, dig!**

Before I close on metal detectors I want to emphasize what you can expect in the way of performance. There are two things to consider should you decide to buy a high-quality metal detector designed strictly for gold. They are its circuitry and the size of the coil being used. The circuitry remains fixed; coil sizes vary. A good rule of thumb: The smaller the coil the smaller the nuggets it will find, but the shallower its effective detection depth even allowing for the halo effect of buried metal. Conversely, the larger the coil the larger the nuggets it will find, and with greater effective detection depth. However, large coils have difficulty "seeing" and pinpointing very small nuggets. Detection depth, *in air*, is generally equivalent to the larger diameter of an elliptical coil or the diameter of a circular coil. Thus, if you use a circular coil of eight inch diameter, its effective detection depth *in air* is approximately equivalent to eight inches *measured downward from the center of the coil.* This is so because the electromagnetic field generated by the coil into the ground shapes like an inverted, elliptical cone. Closely spaced overlapping sweeps of the coil allow for maximum effective depth over a greater area by bringing closer together the centers of the coil's electromagnetic field. Detection depth in the field is usually less than in air due to mineralized ground. Because of these characteristics it is imperative that you use your detector slowly and patiently.

Another way to tackle the upper tributaries and their lack of water is to wait to mine them during the dry seasons. When the gravel has dried out after the winter and summer rains a lightweight, portable drywasher is a good choice for processing the paydirt gravel lying on the bedrock. This device eliminates the need for water to process the gold-bearing gravel. You can clean out the cracks in the way I have already described or you can use a gasoline-powered vacuum cleaner. The ideal combination is to have both items of equipment, with another person along to carry one of the items and you the other. At day's end, you will have processed a respectable amount of gravel and reduced it down to a very high grade of concentrate. The vacuum cleaner, used properly, leaves virtually nothing behind and will get the gold that sweeping will not get. When I go sniping, usually by myself, I take along my vacuum cleaner. Not only is it more efficient than sweeping out the cracks, it is faster. Mine comes with a backpack frame that makes it a breeze to carry in. I usually do not fill it more than two-thirds full of dirt due to the weight of the material, about 50 pounds, which I have to carry out. The total weight of all my tools, vacuum cleaner and dirt is around 70 pounds.

The best tool to use in conjunction with the vacuum cleaner is a long, thin, pointed screwdriver. With the vacuum's engine running on fast speed for maximum suction and the suction nozzle set along the crack, or pocket, I use the screwdriver as a pick to loosen and lift the gold-bearing material. All this material sucks into the nozzle and deposits into the bucket. Methodically picked at, the crack (or pocket) cleans out inch by inch, leaving almost nothing behind. When I mine the upper tributary bedrock in the Las Guijas Mountains, using a combination of drywashing and crevicing, I rip up this bedrock with a sharp-pointed shovel or my pick-mattock. I bring in the vacuum cleaner only after thoroughly sweeping the bedrock and depositing the sweepings into my container. The vacuum cleaner sucks up all the material remaining on the bedrock and I use the screwdriver to pick any cracks and crevices that may contain untouched dirt. After vacuuming, I screen the material in the vacuum bucket through a one-quarter inch mesh screen that removes the larger rocks. Then, I process the remaining screened material through the drywasher in the usual manner.

As you can see, getting gold by sniping doesn't have to be expensive. You can get by with very little capital investment or spend a little more on certain equipment to make your job easier and more profitable. Since you will be working with small amounts of material, quality is what counts. Sniping is a good way to sharpen your eye and improve your gold-hunting techniques.

Above: Yuccas and barrel cactus, Cañada Del Oro Mining District, Pinal County, Arizona.

Fig. 79. A very large boulder sitting on bedrock in an upper tributary of Sucker Gulch, Greaterville Placers. The point of the shovel indicates where the author will start the excavation to reach the gold-bearing gravel behind and under the boulder.

Fig. 80. The author excavating behind the boulder. I have cleared away the top layer of overburden and have reached paydirt. From now on, I will screen all excavated gravel into a plastic bucket.

Fig. 81. Screening the paydirt into a plastic bucket for later panning. Note the circular screen that fits over the mouth of the bucket. You can obtain screens of several mesh openings. Here I am using one-quarter inch mesh. Be sure that you check all your waste screenings for nuggets too large to pass through the mesh.

Fig. 82. The author starting an excavation just behind a bedrock outcrop extending into the wash. Many times there is a small, deep pocket at this spot that traps coarse gold. It will pay you to test this situation.

Fig. 83. Sweeping the bedrock. Make sure that you save all the sweepings as much gold is found in them, gold that you cannot see as it is covered with dust and dirt.

Fig. 84. Vacuuming every crack and crevice to insure that every particle of gold is collected. The gasoline-powered vacuum used here by the author leaves virtually nothing behind.

Fig. 85. The author probing a bedrock crevice to remove the compacted gold-bearing dirt that has accumulated in it. Use a long, pointed screwdriver or a crevice tool to loosen the dirt. Recover the gold-bearing dirt with a spoon, or use a gasoline-powered vacuum to suck it up.

Fig. 86. The author metal-detecting bedrock exposures in an upper tributary of Sucker Gulch, Greaterville Placers. Coarse gold nuggets frequently become trapped in bedrock cracks in these upper tributaries and it will pay you to metal detect them.

CHAPTER III

HIGH BANKING - YOU CAN BANK ON IT

In the Introduction I touched on a gold mining device that most of today's prospectors tend to ignore, the high banker. Perhaps one of the reasons is that high bankers require water available in sufficient quantity for their operation. Their use, limited to areas of reliable stream flow, restricts them to a few localities in Arizona. Another is that, where water is available in gold-bearing areas, many prospectors believe that gold occurs primarily in stream beds. They concentrate their efforts toward getting only that gold, to the exclusion of other possibilities. In a sense, they are correct about their assumption. Ultimately, gold will find its way in the streamcourses and into the stream bed gravel. Eventually, it works its way downward to bedrock. The one problem with going for the gold in the stream beds is that almost every other prospector is after it, too.

In popular gold placers such as Greaterville, miners have worked the stream beds continuously ever since the establishment of the district in 1874. In the depression years of the 1930s, unemployed men in the hundreds worked the stream bed placers and, today, scores of recreational miners continue to rework the gravel. For the most part, the gold recovered today is gold newly eroded out of the hillsides or gold re-concentrated from other places overlooked by previous miners. This is not to say that you can't get gold from the stream beds in heavily worked over placers. It's just that your take will be modest and the size of the gold smaller. It becomes a matter of finding a spot previously untouched or one not worked for many, many years, which is very hard to do in popular areas.

Our success with working the stream gravel of Lutz and Ash Creeks on our former claims in the Huachuca Mountains was a result of three factors. They are (1) the remoteness of the placer; (2) it was relatively unknown to most prospectors; and (3) the stream bed was relatively untouched since the 1930s. Although one lone prospector lived there on a life lease from the Forest Service, he hadn't done any mining in decades. Those three factors were the major ones contributing to our success in finding very good gold, gold that was in quantity and very large. If miners worked that area as much as Greaterville I doubt that we would have enjoyed the returns we got. Since we had water available in the creek sufficient to operate suction dredges, at least for five of the ten years we owned the claims, we concentrated on mining this way. Suction dredging is one of the best ways to get gold because of the tremendous amount of material you can move in a day, but what do you do if miners with suction dredges work the stream bed year after year? You could do high banking.

We were lucky in that we had ground relatively untouched for decades and freedom of choice. It's very hard to leave gold you know is there in situations such as ours, but I did on a regular basis. Ash Canyon in the Huachucas has many high banks, remnants from when the creek flowed in a different position for thousands of years, or as a result of down cutting. Let's go back several millions of years and try to imagine what the area looked like. The gulches we see now probably didn't exist. Mineralized intrusions had forced their way upwards through the host rock and gold-bearing solutions worked their way along cracks and fractures in the rock. As the solutions cooled, the gold precipitated out along the selvages of the seams, cracks and fractures. The forces of nature uplifted the mountains along with the intrusions and erosion immediately set upon them. Over the millennia the forces of erosion wore away the rock cover, exposing the gold-bearing veins, stringers and ledges and releasing the gold from the grip of the mother rock. The gold, freed from the rock, washed down the slopes and into the watercourses. The watercourses, for the most part, followed natural faults in the rock. For thousands and upon thousands of years the gold continued to erode and wash down and into the watercourses. The watercourses carried this gold and deposited it into the favorable spots where the water slowed. Then, the watercourses shifted, perhaps suddenly during catastrophic flood conditions that caused landslides that blocked the channels; or, the watercourses cut downward below their banks. In their new positions, the watercourses continued to flow and bring down gold. However, they left the earlier gold high, and dry, that they had brought down previously. Now virtually locked in position, it will take many millennia for this gold to erode out. This is the gold in the high banks. This is the gold for the high banker!

The gold in the high banks may be in remnant bends and gravel bars, behind and under large obstructions such as car-sized boulders, and in ancient stream channels. For the most part, this gold occurs in streaks, that is, not distributed evenly throughout the high bank. We are going to have to read the high banks before we begin any mining. First, we have to determine just what is a high bank. If you walk along the present watercourse and look upward from the stream channel, you might see a flat-topped berm of gravel and dirt paralleling the watercourse. It may be anywhere from ten or more feet above the channel. There may be several of these berms, or terraces. This is a high bank. It could extend for many hundreds of yards along the stream or for only a few, cover many acres in extent in places where stream gradients lessen or be very limited in area. In some cases, it is not easily discernible and tons of recent debris cover it. You will have to look hard to see the slight clues to its existence.

Some of the things that you want to look for are (1) rounded cobbles and pebbles that are stratified, that is, the smaller ones laid down on top of increasingly larger ones (in some placers the gravel may be angular, indicating they haven't travelled far); (2) brownish-stained streaks in exposed side banks or hillsides; these indicate concentrations of heavy minerals, mainly magnetite and hematite, the "black sands" conditions favorable for the deposition of gold; (3) whitish to rusty-stained quartz pebbles in lone layers that stand out from the rest of the gravel; (4) a combination of the above in localized areas or seen as narrow bands descending the hillsides, side banks and across alluvial fans; (5) gravel layers distinct from the predominant material, particularly rounded gravel in areas covered mostly with angular material, which may indicate an ancient stream channel; and (6) a "clip."

A clip is a localized concentration of rounded boulders and rocks in a gulch, that ends abruptly upstream. If accompanied by much black sand and rounded quartz pebbles, it may indicate an older stream channel that is eroding out of the hillside. It may be traceable to its source by following the path of these materials to where they end abruptly. Not all ancient channels are rich in gold, but in an area noted for its gold placers it will pay you to investigate each one. You will have to dig down into the ancient channel to its bottom where most of the gold concentrates.

You may see trees on eroded high banks with their roots exposed. On a geological time scale, trees are recent arrivals. Their ages hardly exceed a few hundred years. The gold found in the dirt surrounding the tree roots was probably there already and the tree roots grew around it; then, the roots became exposed as the high bank eroded during flood conditions. So, tree roots are not necessarily as important in this situation as in the stream bed. What is more important are the larger boulders. These are not recent arrivals; they've been sitting in the high banks for hundreds of thousands of years. You should take out the gravel behind and under these for the gold that deposited here long ago, **but be extremely careful!**

When you do high banking, you are going for gold deposited over millennia. Relatively untouched for thousands and perhaps millions of years, these situations may be very rich in gold. Some of the ancient stream channels may have gold re-concentrated from even earlier stream channels and you could find yourself a bonanza. Gold in the high banks deposited in an earlier time and may not have transported very far. This gold is usually coarse and ragged in Arizona.

The high banker is a device that consists of sluice box and riffles; header box with a grizzly (a screen that classifies the material put through the sluice box by taking off the larger, waste rocks); a spray bar or system of delivering water to the header box to wash the gravel; a stand, or legs, on which the sluice box sits; a length of pressure hose for the delivery of water to the header box; gasoline-powered pump, and foot valve. The sluice box usually has matting made of a carpet-like material upon which the riffles set and a classifier screen above the riffles that may extend into the header box. The high banker *must* have ample water for its operation. Depending upon the size of the high banker, if you can operate a suction dredge in the stream or creek, then you can operate a high banker. Usually, you do not save the tailings water from the sluice and recycle it. If it is necessary to do so, build yourself an impoundment for the foot valve and channel the tailings water back to this impoundment. Due to the amount of water necessary to operate a high banker, it is not practicable to carry in your water and recycle it as is done with a mini sluice. At a minimum, you would need 40 gallons of water for the smallest high banker, and that poses a problem of weight, and of a vehicle capable of transporting such a large volume of water to remote areas.

The pressure hose for delivery of water to the high banker's header must be long enough for you to reach the high bank from the foot valve. The reason for this is that you will set the high banker up at the high bank and not at the stream. A length of pressure hose of about 50 feet is adequate in most cases. However, when I high-banked Lynx Creek in the 1970s, I used a length of hose of 100 feet to reach the place where I had set the high banker. In this situation, I modified my suction dredge's pump to accommodate a hose of 3/4 inch diameter and used my dredge's sluice box, also modified, to accept the hose to perform this method of mining. From what I learned here, I was able to go on to design and build my own high bankers. Now, I use a high banker supplied by two 50-foot lengths of 3/4 inch garden hose. The water pump, attached to my 5-hp gasoline engine, has a manifold made of PVC piping that I built to accommodate the two hoses.

You can buy a ready-made high banker from your local prospecting dealer. As this is a highly specialized and quite expensive item, you might consider a dual-purpose one suitable for suction dredging. The pressure hose that comes with these units is much too short and you should consider buying one at least 25 feet long.

Let's say that you found a promising spot and want to do some high banking. Set up your high banker so that the header box faces the bank and the sluice points to the stream. Attach your hoses, foot valve, etc. Fill the gas tank of the engine and check the oil level in the crankcase. Prime the pump and start the engine. Adjust the speed of the engine so that you have a full flow of water through the sluice, but not too fast. Adjust the pitch of the sluice box to about one inch per foot and level it so that the flow of water is even through the sluice box, not more to one side or the other.

The high banker is a hand-fed device requiring you to mine the gravel with a shovel, pick and other hand tools. You feed the header continuously with shovels full of material and let the grizzly do the classifying for you. If you pre-screen the material into buckets and empty these into the header, it will slow you down. The key is quantity. The design of high bankers is for volume, but you must ensure quality, also. The high bank material will have gold in streaks. Therefore, you must cut the high bank vertically from the bottom of the deposit to the top. I don't mean the entire high bank, only that portion exhibiting potential such as the rounded boulder channels, etc. What you want to do is find that paystreak or layer with the most gold and put that through your high banker.

In California during the gold rush of the 1850s through 1870s, miners used hydraulic methods to mine the high banks. Pipes, carrying water from streams higher up in the mountains, connected to nozzles. These nozzles, or "monitors," directed streams of high pressure water at the high banks and washed out the gold-bearing gravel. Sluices, set up to intercept the washed out gravel, recovered the gold. This mining method caused monumental environmental damage. A court order finally stopped it. After more than 100 years since stopping, the scars from this type of mining remain today.

It may tempt you to "hydraulic" the high bank by directing at it a stream of high pressure water from the pressure hose of your high banker. This is an absolute no-no! Federal regulations strictly prohibit hydraulic mining on public lands and in our national forests. The penalties for doing this are severe. Don't even *think* about doing it.

Since the BLM and the Forest Service permit you to do mining by hand methods you should start as follows. Beginning at the bottom of the layer and using your shovel, excavate a cut about three inches deep, one to two feet wide and one to two feet high into the bank to expose the virgin gravel. Now take out the material you exposed to a depth of a couple of inches and process this through your high banker. Shut down the pump motor and pan out the concentrate in the sluice box and header screen. Check for gold. Restart the pump engine and take out another section of bank higher up and run this material through the high banker. Pan out the concentrate as before. Repeat the process until you have cut vertically through the section of high bank you think carries gold. From each panning, you should be able to determine which portion of the high bank channel carries the most gold or has the highest potential for carrying the most gold. If the material you are working is an ancient stream channel, you might consider the lowest layer the one with the highest potential and put only this section through your high banker. If you have the time, energy and resources (plenty of water available and gasoline for your pump motor) and the channel you are working is small, you may want to put the entire channel through your high banker. This will be much work but you will get at all the gold in the channel.

Once you have determined the paystreak layer, you should excavate it entirely and put it through your high banker. The operation should be one of continuous feed for maximum return in a minimum of time. Therefore, you should be able to take out the high bank material directly with your shovel and put the material through the header in one continuous motion. If your high bank material contains many large boulders, this will slow you down or hinder your ability to feed the high banker continuously. You will have to excavate the gold-bearing dirt from between the boulders with pick, trowel, shovel and pry bar and put the material into plastic five gallon buckets. In this case, you should have on hand at least six or more buckets and fill these to capacity before you turn on the pump to the sluice. Pour the contents of the buckets slowly into the header. Empty each bucket in turn. After all are empty, you shut off the pump, and empty the contents of the sluice into a reserve bucket for later panning; or, refill the six buckets with more material and empty these into the high banker. You may repeat the steps over several times before you empty the sluice concentrate into the reserve bucket.

Some high banks are not high, and the overburden covering the paydirt may be only a few feet, with blue sky above. Here, you will have to remove the overburden to continue mining. However, in very high banks several feet of overburden may cover the paydirt. In this case, you may only be able to follow the paydirt laterally. To do otherwise means that you will excavate a tunnel into the hillside. This is not dangerous as long as you can remain outside the opening and reach into it with your shovel. It is very dangerous otherwise and particularly so if you come across very large boulders. A cave-in can occur without warning, in the wink of an eye, severely injuring or killing you. Many miners have died following the paydirt into the hillsides; and, some, never found again after a cave-in buried them and completely obscured their diggings.

When I find myself digging an opening into the hillside, I try to collapse the paydirt to the floor of the opening. I do this by using a sharp-pointed steel pry bar or the point of my shovel. The top of the opening is arched upward from back to front (higher at the front of the opening) and at each side. This provides stability to the overlying dirt and prevents a cave-in. The collapsed paydirt is easily scooped up from the floor with the shovel.

I have seen many situations where miners are working around a boulder of several tons that is protruding from the high bank. Although you can find good gold in this situation, remember that you do not have any idea of the size and extent of the boulder in the hillside. Consider what is holding it in place and the stability of the material around it. It can let loose in an instant, particularly, after heavy winter rains have saturated the ground. If you are lucky, you may see the ground above it crack open, signaling an imminent collapse. Then you can get out of the way. More often, mining absorbs your attention and you forget the danger above. My advice, be very, very careful. In the dry seasons you may be more fortunate in that the dry ground is as if cement, and it holds the boulder in place, longer, but not permanently.

When I do high banking, I am very careful around car-sized boulders. Some boulders are much smaller and weigh only several hundred pounds. I feel much better if I can collapse these beforehand, taking extreme care to be out of their way when they fall. To collapse these, I use a hexagonal, steel pry bar of five feet in length to pry the boulder loose. I stand off to the side, uphill of the boulder and work it loose with the bar, making sure that it will fall clearly away from my person and equipment.

You also can use your high banker a little differently than that just described. Instead of searching out the ancient stream channels you can concentrate on bedrock outcrops above the present watercourses. Here, you can mine the gravel of bedrock cracks and pockets, and going after gold deposited during times of flood run-off. In a way, this is like sniping, but, because you have water available from the stream or creek you are able to mine on a larger scale.

If the high bank consists of a terrace of gravel on bedrock, a condition that occurs frequently in many of the placer areas, particularly in the Bradshaw Mountains, you should mine the top several inches sitting on bedrock. If the amount of overburden isn't excessive, you can tackle the terrace from the side where the bedrock exposes, or you can dig downward from the top of the terrace to the bedrock. The gold in this terrace may consist of recent, and ancient, deposition. As a rule, the higher the terrace is above the present stream bed, the older it is.

I would like to discuss one last point before I leave the subject of high banks and that is caliche (pronounced "cuh-lee-chee"). In some placer areas, Ash Canyon in the Huachuca Mountains as an example, caliche, or carbonate of lime, cements much of the high bank gravel. This makes high banking a difficult proposition as the dry gold-bearing material has the hardness of concrete. Although the gold found in these situations is coarse, getting it is much work. You may want to consider trying various spots in the high banks that are free of caliche to avoid hard labor even if it means leaving some gold behind. This is a decision you must make on your own.

If you have water available in sufficient quantity high banking is a good way to get gold that others bypass or simply ignore. It is mining on a larger scale than sniping or micro-sluicing. In an area overworked by suction dredges, this may be the way to go. You are going for gold that may give you some pleasant surprises. You can bank on it!

***

Fig. 87. The author's high-banker that he designed and built. The unit folds down to conserve space. A manifold attached to the pump delivers water under pressure through two hoses attached to the header of the high-banker. The header box is removable.

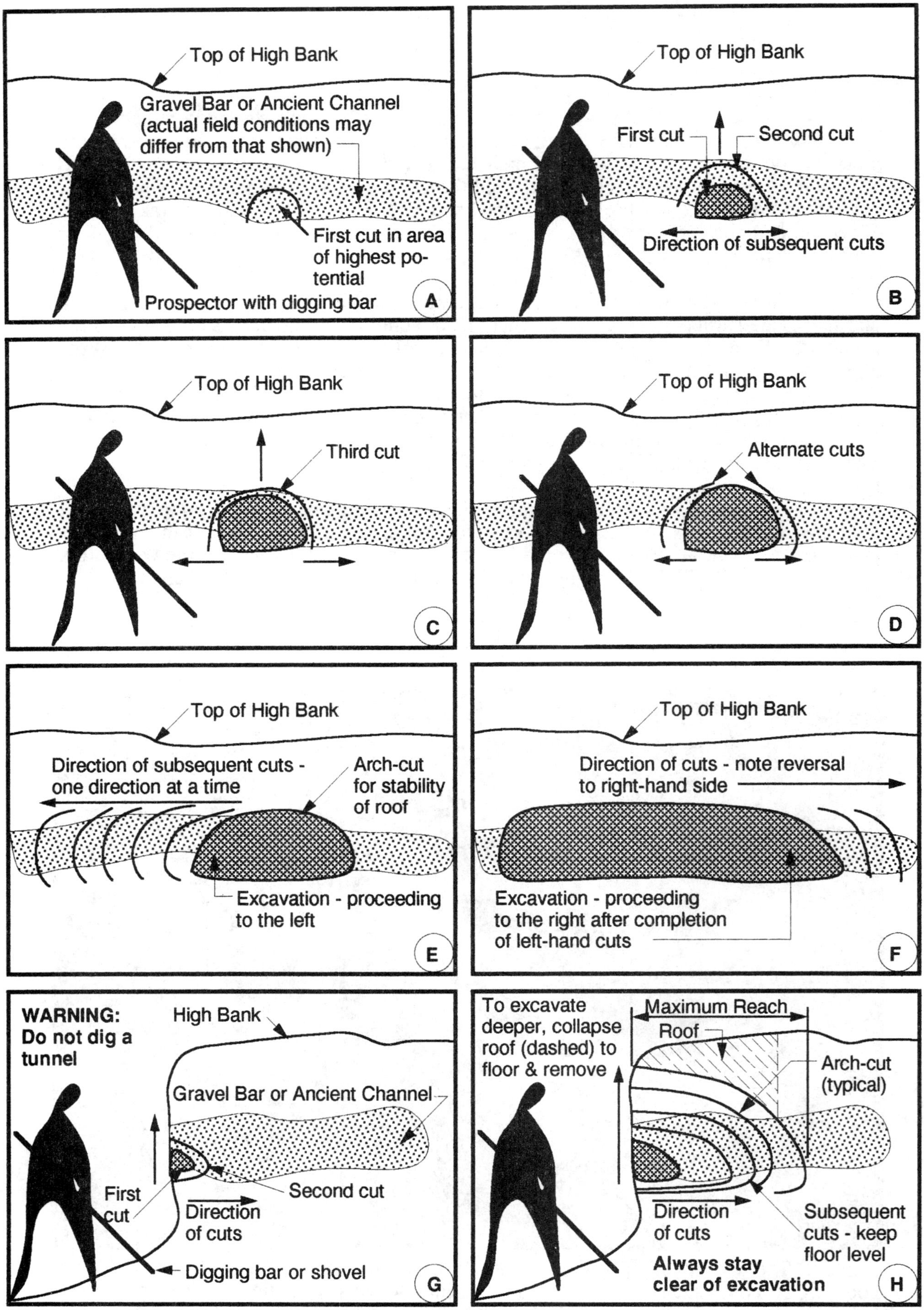

Fig. 88. **EXCAVATING A HIGH BANK GRAVEL BAR OR ANCIENT CHANNEL**

Fig. 89. The author's high bank set-up at a bedrock outcrop in Ash Canyon, Huachuca Mountains, April 1993.

Fig. 90. The author feeding the hopper of his high banker. I have set the high banker so that I can feed it in one continuous motion with each shovel full I dig. The gold I recovered here ranged up to 1/20 inch in diameter.

Fig. 91. Two recreational miners from Tucson using a high banker at the well, Ophir Gulch, Greaterville Placers. On the left, Larry Walker; at right, Don Lindberg. Note the sluice extension for fine gold recovery. The bench excavation is out of the photo to the right. Photograph used with permission of the subjects.

Fig. 92. Larry Walker tending the high banker. He is washing the bank run material over the grizzly, protecting his hands with heavy-duty rubber gloves. Note the tailings pile showing that the bank run material is not pre-screened. Photo used with permission.

Fig. 93. The remains of an ancient gravel bar now exposed by miners, Ophir Gulch, Greaterville Placers. This may also be an ancient channel extending into the hillside. Note the rounded cobblestones. Because there is no stratification of the cobblestones, I believe that they were deposited during flood conditions. The bar probably carries coarse gold unevenly distributed in erratic streaks. This is not destructive mining.

Fig. 94. An example of destructive high-banking. Gold miners have torn into the roots of an oak tree in their fever to get at gold deposited millennia before the tree took root. This way of mining is a sure way to have the Forest Service close an area to prospectors.

CHAPTER IV

MINI OR MICRO - A SLUICE BY ANY OTHER NAME

There is a gold recovery device gaining widespread popularity among gold prospectors in Arizona. This is the mini, or micro, sluice.

Here is the baby brother to the high banker, but with one exception. It doesn't rely on having water available in the creek, or stream, for its operation. No, it is not a drywasher. What it does is it brings along its own water; or, rather, you do.

This little device is almost the perfect answer for getting gold out of Arizona's placer areas. Whether the gold placer has water available, you can operate this equipment and get the gold the best way, with water.

Water is the best way to recover gold based on one simple fact. Gold is about 19 times heavier than an equal volume of water, but only about 6 times heavier than the dry gravel in which it occurs. Although other factors, such as the alloy content of the gold and the specific gravity of the heavy minerals associated with gold, may alter these figures somewhat, the use of water remains the best way to separate gold. Water virtually "washes" the gravel, freeing gold from clay or dirt clods. The use of air in a drywasher will not free the gold in these situations.

Whereas a mini sluice operates whether the placer is dry or not, a drywasher does not. If the gravel is moist, it spells trouble ahead for the drywasher, and if very moist or saturated with water, death. Some of Arizona's best gold placers are dry, most of the time. However, rain does fall and these areas do get wet. Winter rains pose the worst possible scenario for the drywasher. Coupled with lower air temperatures, shorter days and diminished sunlight, it results in the ground remaining wet for weeks and unworkable with a drywasher. Not so for a mini sluice, it works wet or dry gravel.

It sounds as if the mini sluice is the perfect answer to mining Arizona's gold placers. Well, almost. As I said, the mini sluice uses water, and in an area without water you bring it along for it. Now, mini sluices come in a variety of designs and sizes and all depend on water, but the amount of water varies with the design and size of the unit. In the dry areas, the mini sluice sets up to use water that recycles. Essentially, two vessels connected to each other by an overflow pipe (not unlike the principle of a septic tank) contain the water. The smaller mini sluices use two five gallon plastic buckets for the vessels and the pipe is a made of PVC plastic. Some of the larger mini sluices use two, 20 gallon, galvanized steel garbage containers to hold the water. A PVC plastic pipe connects the two. Since individuals build most of the mini sluices, there is a great variety available to the gold prospector. Many are obtainable from your local prospecting dealer, either as newly built, or second hand, equipment.

As I said, they are almost the perfect answer for Arizona's gold placers. However, here is something to consider. If you cannot obtain water nearby where you plan to mine, you must bring it in with you. Water is very heavy. The weight of ten gallons of water is approximately 86 pounds. Allowing for losses in the field, you probably will bring in double that amount, or 20 gallons. Now, the weight jumps up to around 172 pounds. I have seen larger mini sluices that use double the last amount, or 40 gallons! These amounts are only enough for one day's work. There are miners with mini sluices that bring water in 55 gallon drums to have on hand a sufficient supply. That is the equivalent of almost 475 pounds! **With equipment that uses water, you can never have too much of it.**

So far, I have been talking only about the water the equipment uses. Prospectors need water, too. They also need food, tools and a variety of other things. Everything they bring in, including themselves, has weight. Something must transport this weight, and that something is a vehicle capable of carrying it. So, the other consideration is the vehicle. A passenger car is not that kind of vehicle. What you will need, then, is a utility vehicle with springs capable of carrying the load. The vehicle may be a sport utility vehicle, or a pickup truck. If the area in which you will prospect is remote, with very rough roads, a four-wheel-drive vehicle may be the only answer. From a mini sluice to a four-wheel-drive vehicle to carry it is a major jump in cost. If you already have that kind of vehicle, the cost of a mini sluice is relatively small. Mini sluices vary in price from $400 and up, depending on whether they are new or not and the features each have.

Finally, where to put the mini sluice to mine the gold relates directly to the water-weight problem. If you are in a dry placer area, you want to get as close to the gold-bearing location with your vehicle as you can to set up your mini sluice. Otherwise, you will have to lug the water to the equipment set up some distance away. It is not much of an advantage to lug the dirt to the equipment as dirt weighs more than 100 pounds per cubic foot. To get an idea of this weight, fill a five gallon plastic bucket with dirt to within an inch or two of the rim. Now, lift it. This is *only* about 70 pounds! Imagine yourself carrying this bucket 50 to 100 feet, and possibly up hill, to your equipment. After several trips of this, not counting the work you already put in mining that dirt, I would almost guarantee that mining for you is no longer fun but very hard work. Even if you don't mind carrying the dirt, the time consumed in transporting it will slow down your operation. What *is* the answer? You may not build roads to your operations without a clearance from the governing agency. Therefore, you may have to content yourself with working the more accessible, and popular, spots; or, resign yourself to carrying the water, or the dirt, a *short* distance. There are no easy answers.

I mentioned previously that there are as many designs, and sizes, of mini sluices as there are individuals making them. There are a few commercially made. All are smaller versions of a high banker, and some come with innovations such as a jigging action to settle the gold. The units themselves are very light weight. Some sluices come with grizzlies to screen off the larger rocks. The sluice box is narrow, roughly six to eight inches wide, and approximately three feet in length, and the riffles may be anything from flat steel strips set at an angle, to round rods, or expanded metal lath. Most come with some type of matting to catch the fine gold. Water delivers to the header through a pipe or spray bar to which connects a length of hose attached to an electric powered water pump. This pump operates off a 12 volt car, or motorcycle, battery. The water pump may attach directly to one of the plastic buckets that's filled with water or simply dropped into the bucket. An overflow pipe made of a short length of PVC piping connects the buckets just below their rims. The spacing of the buckets is enough to accommodate the length of the mini sluice. The mini sluice may attach directly onto the buckets or stand on legs that elevate it above the buckets. Water fills both buckets above the overflow pipe. The bucket at the sluice end catches the waste water, and tailings that settle down to the bottom of the bucket. A circular screen fits into the top of the bucket to catch the coarser gravel coming off the end of the sluice. You must empty the screen frequently during your operation and check it out carefully for nuggets before you toss away the waste material. The pump, set into the other bucket, provides settled-out water to the sluice. In some designs, a valve controls the amount of water delivered to the sluice. After a day's operation, the tailings usually fill the first bucket up to the overflow pipe and the water is very muddy. To go on you must empty both buckets and refill them with fresh water. Without sufficient water on hand for refilling, your operations limit to one day's work.

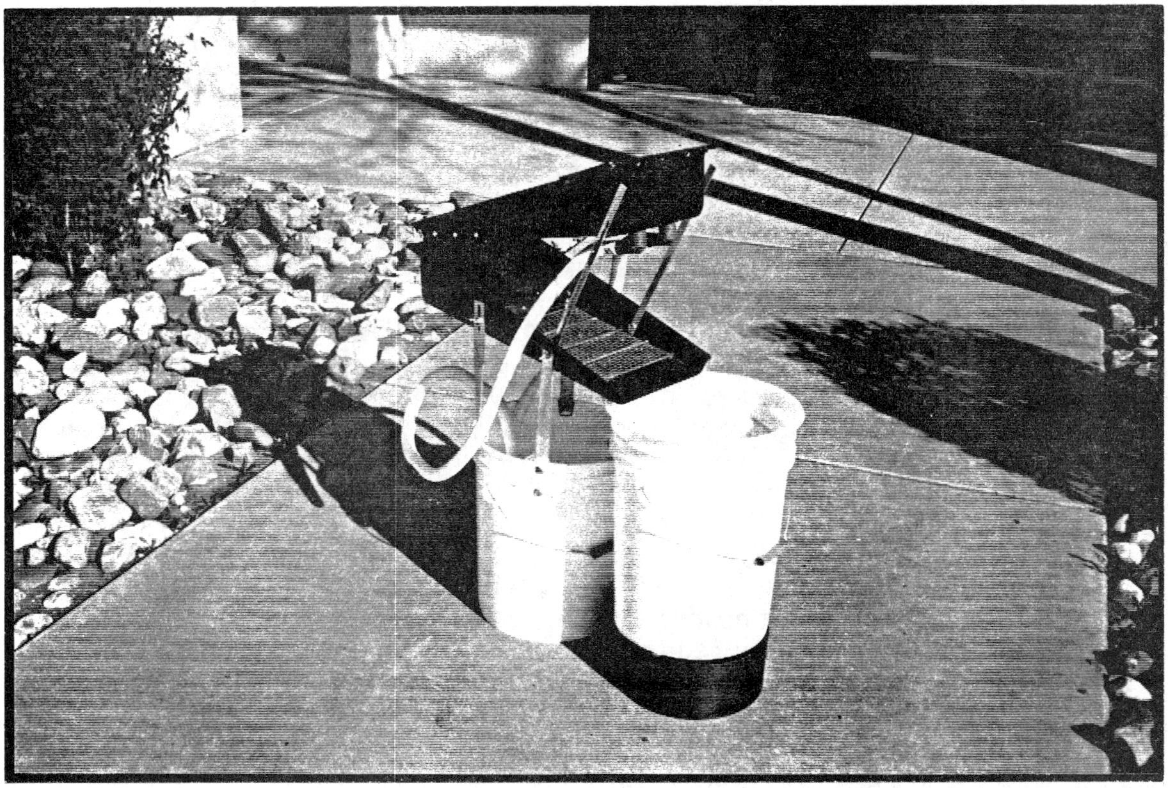

Fig. 95. An example of a mini sluice using plastic five-gallon buckets to contain recyclable water. The bucket under the sluice end intercepts the tailings and also functions as a settling tank. An overflow spout allows settled water to return to the other bucket that contains the pump. Mini sluice courtesy of Desert Trails, Tucson, Arizona.

The sluice box inclines to the tailings bucket at about one inch per foot. A pitch adjustment permits you to have the sluice set at the proper angle for the material passing through it. You must classify the material put through the sluice, that is, screen it to a size that permits easy passage through the sluice box and efficient recovery of the gold. A grizzly that is a part of the sluice performs this classification, or, you do by pre-screening the gravel into a bucket. Due to the small size of a mini sluice and low water flow, I do not recommend screenings over one-half inch in size. One-quarter inch size and smaller is better. This limits your gold recovery to particles passing through the classification screen. If losing larger nuggets concerns you, consider taking along a metal detector and checking the waste screenings.

Because the mini sluice is not a high volume piece of equipment, material fed into it should be high grade. This is the material excavated from cracks, fissures and pockets in bedrock or the paydirt occupying the first few inches above it. Don't try to put the entire stream bed through the mini sluice. Whether you feed the mini sluice with a shovel or with a bucket you must do it slowly. This prevents overloading the header and the riffles. All the material fed into the sluice should wet thoroughly and break down to release the gold.

The mini sluice makes an excellent tool for testing the stream bed and side banks for paystreaks. In a sense, it is like a giant gold pan processing a very large portion of gravel from which you can determine the gold content quickly. How much gold you get depends on many factors not the least of which is the richness of the material processed. However, if the gold is there the mini sluice will get it. Like all sluices, you must watch for overloading of the riffles with black sand. If the riffles fill up with black sand, gold will run off with the tailings; it just won't be able to settle out.

Apart from the water-weight factor, mini sluices are wonderful devices for getting gold in dry situations. They are versatile, portable and light in weight. They are almost the perfect answer to an Arizona gold prospector's prayer.

***

Fig. 96. The author, at left, and Mike London, at right, processing paydirt using a mini sluice with self-contained, recyclable water. Peppersauce Canyon, Santa Catalina Mountains, April 1993. Due to their small size, portability, and the ability to use self-contained, recycled water, mini sluices make excellent devices to take along to Arizona's dry placer fields. Photograph used with permission.

Fig. 97. The author, at left, feeding pre-screened paydirt into the grizzly, and on the right, Mike London spreads the material to assure thorough washing by the spray bar. Photo used with permission. Mini sluice courtesy of Discount Prospecting Outfitters, Tucson, Arizona.

Fig. 98. Taking a break after a few hours of "hard" work mining gold in Peppersauce Canyon, Santa Catalina Mountains, April 1993. From left to right, Debra Torres, Alyn Wielgus, the author's wife, and Mike London. Photo used with permission.

CHAPTER V

THE HAND THAT ROCKS THE CRADLE

There are many of you starting out in gold prospecting who don't have the funds available to invest in the more productive gold recovery devices. However, you would like to have a piece of equipment capable of handling more gravel than a gold pan. Perhaps the answer for you is a rocker box, or cradle.

I feature this device in a sketch I drew for my book ***Arizona's Golden Secret*** that shows an old-timer from around the turn of the century rocking the cradle. He may not have ruled the world but I think that he was king in many ways. Beholden to few, he was fiercely independent, ruggedly individualistic, self sufficient, hard working, honest and smart. He could not be otherwise in the environment he found himself in. To win the gold and survive in hostile terrain required that he possess the strength of character we all admire. In his hand is the simple machine that enabled him to go on. By the time he was using it, it already had a long history of use in the gold fields. Surprisingly, with all the new development in gold recovery devices since, this machine remains an old standby among many prospectors. Its longevity is attributable to three factors (1) it is simple; (2) it is inexpensive to make, or buy; and (3) it gets the gold.

The cradle, or rocker, uses water. Using water is the best way to recover gold. One very good feature about the cradle is that it uses very little water and is suitable to take to the dry placers. Many an early prospector using a cradle constructed an impoundment, or settling basin, to contain the water that he poured into the hopper of his machine. He usually packed in his water on mule back in the dry seasons. He set up his cradle so that the tailings did not settle directly into his water impoundment. The water coming off the end of the sluice channeled back to the impoundment, allowing the silt to settle out. If he was working a flowing creek or stream, all the water he needed was at hand and an impoundment was unnecessary.

For the moment, I am assuming that if you acquire a cradle, you will work in an area without available water. Therefore, you must take along the necessary amount of water to operate this machine. Ten gallons of water is the minimum amount you should take with you. You must recycle this water. A twenty-gallon galvanized metal tub is the container you should use to contain the water. Your cradle sets up higher than the tub so that the sluice end is over the tub. Tailings, and muddy water, discharge directly into the tub. As the tub fills with tailings the water level in the tub rises. The lighter material settles out as a fine, silty mud. From time to time you remove the tailings in the tub using your cupped bare hands with the fingers slightly apart. Let the water drain out between your fingers and into the tub before you toss the waste away. You must conserve as much of your precious water as possible. Using a kitchen strainer, or even a flour-sifting scoop with the handle removed, is a very good way to remove the waste material because of its small mesh openings.

As you use the water it becomes very muddy. This has no bad effect on recovering the gold so long as it doesn't become a mud slurry. Some placers are notorious for a high clay content in the gravel; these will muddy up the water very quickly and turn it into a slime before the day is out. The only thing you can do is allow it to settle out overnight.

Your first day is your best day. From then on, the water available becomes less, and muddier. However, if you allow the water to settle out overnight it will clear for you. To prevent undue water loss through evaporation, cover the unused tub with a sheet of plastic or a tarpaulin. Using these measures, you will conserve enough water for at least two full days of operation, and possibly three.

Don't expect to process anywhere near the volume of material that the other devices do. A cradle is hand fed and hand operated. By its very nature it is slower.

Before you begin, you should already have mined the paydirt that you will process through the cradle. Have this paydirt in plastic five gallon buckets, close at hand and ready to go. Have something to sit on while you work the cradle; an upturned five gallon plastic bucket makes a good seat. You need a long-handled dipper to scoop up the water that you pour into the hopper. Fabricate this from a one-pound coffee can from which you remove the cover. Attach a short length of broom handle to the can by either pushing it through holes in the upper end of the can, or by wiring it to the open end of the can.

You can feed the dirt into the hopper using a garden trowel, small shovel, or a metal scoop. The left hand grasps the handle attached to the cradle and rocks it. The right hand feeds the paydirt and pours the water into the hopper.

The cradle, built of wood, consists of a removable hopper; removable apron, and a sluice. The hopper is a rectangular frame usually made of one-by-fours with a classifying screen attached to the bottom. Galvanized wire mesh of one-quarter inch size makes up the screen. The hopper frame is approximately an inch smaller than the width of the cradle. It rides on two cleats. As you rock the cradle from side to side, the hopper slides along the cleats and strikes each side of the cradle. The jarring of the material in the hopper by this striking helps to settle the gold that drops through the screen and onto the apron.

The apron, usually made of a heavy cloth such as canvas, inclines at about a 30 degree angle under the hopper. There is a sag in the cloth at the lower end. This sag helps to catch the heavier gold. As the paydirt in the hopper processes, screenings and gold drop onto the canvas. Water poured into the hopper falls onto the apron and washes the material, leaving black sand and gold behind and carrying off the waste. Some of the fine gold washes off with the waste. To capture this gold, one or more riffles set across the sluice. Instead of riffles, some cradles have a removable carpet-like matting placed on the bottom of the sluice.

In use, the sluice inclines more sharply than other sluices that carry more water through them. Set up the sluice so that the waste passes through and not builds up on the bottom. You will have to experiment for the best possible pitch. Attached to the underside, and crosswise of the sluice at each end, is a strip of wood with a curved bottom, the rocker. To prevent creeping of the sluice, a nail protrudes downward from the center of each rocker and into a hole drilled in a crosswise wooden plank. As you move the handle attached to the cradle back and forth, the cradle rocks on these two planks.

To operate the cradle, load the hopper only half full with paydirt. Pour out one full can of water over the paydirt and rock the cradle back and forth a couple of times. Repeat the pouring and rocking process. Stop when the material in the hopper is clean and all the smaller particles have passed through the mesh. Pour out one more can of water, rock twice, and check the waste remaining on the screen for any nuggets too large to pass through. Spread out the material on the screen so that the gravel doesn't hide any nuggets. Any nuggets on the screen will stand out because the dirt covering them has washed away. This is the beauty of washing gravel with water. The gold gets cleaned. To be on the safe side, dump all the hopper waste onto one pile. If you have a metal detector with you, check out this pile at the end of the day. Follow the directions I gave in my first book on metal detecting drywasher piles.

After the day's work, remove the apron and check it for larger nuggets. Then, dump out the concentrate remaining on the apron into your gold pan. You may pan out this concentrate now, or save it for later panning. Flush out the concentrate in the sluice into an empty plastic bucket. Save this concentrate for later panning and gold separation.

Many cradles are home made by individuals following their own ideas or plans available in books on gold prospecting. The Arizona Bureau of Mines Bulletin Number 168 gives plans on constructing a cradle. Your local prospecting equipment dealer may have on hand used cradles of several different designs and sizes. He also has access to commercially manufactured cradles if you want yours new. A cradle is a step up from the gold pan. You may not get rich using one but you will get gold, much more, and faster, than with a pan alone.

***

Fig. 99. A cradle, or rocker box, built by Bruce Quick of Tucson, Arizona, courtesy of Discount Prospecting Outfitters, Tucson, Arizona.

CHAPTER VI

BUILD YOUR OWN DRYWASHER

In *Arizona's Golden Secret* I described the one item of equipment every serious prospector should have to test the dry placers - a drywasher. Because Arizona is a dry state, most of the time, a drywasher is indispensable to the prospector wanting to process the dry gold-bearing gravel in remote areas. Although he may opt to use a mini sluice, the lack of water in these areas requires that the prospector bring it along to work his sluice. The problem of the weight of water is no small consideration. It compounds if he wants to work his sluice for more than one day as he must empty the gravel-loaded bucket, and muddy water, to continue working. This requires him to bring along a sufficient amount of water to accommodate several-days' work. This is not a problem with a drywasher.

If the gravel is dry, a lightweight drywasher is ideal to work it. The only drawback with a drywasher is that the material processed through it must be dry. Moist or wet gravel spells death for a drywasher; it just won't work properly, or at all. One reason for this is that moist dirt clogs the pores of the cloth covering the riffle tray preventing the blowing action through the gravel so necessary for separating the gold. The other is that gold clings to moist gravel or locks up in dirt or clay balls and goes off with the tailings. So, even if you *could* puff the moist gravel with the drywasher you may not recover the gold. In spite of this major drawback, a drywasher, when used properly under the right conditions, is a wonderful device for getting gold. It isn't as efficient as a device using water for separation, but it comes very close. This difference may not matter when you consider the alternatives.

One factor in favor of a drywasher is its light weight. Drywashers are available so light in weight that you can carry them to the spots where you work. Since the weight of equipment and tools is a major consideration, the lighter the drywasher is, the better. The lightest drywashers are those that rely on muscle power. These machines are also the simplest to operate. As the machines become more complex and reliant on an external source of power other than human muscle, they get heavier. The tradeoff is the power to process gravel. Lightweight, manually powered machines just don't process the volume of gravel that electric, or gasoline, powered ones do. However, there is one compensating feature with the manual, lightweight drywasher, and that, is <u>simplicity</u>. There is so little to go wrong with it. In remote areas, with virtually no services available for miles, you want a machine that won't give you trouble. This is not to say that the other, powered, machines are a bad choice, only that they are more complex and require more maintenance to prevent malfunctions. A gasoline-powered engine is particularly vulnerable to the dust created by a drywasher in action. Change the crankcase oil frequently and clean the air filter often.

After I published *Arizona's Golden Secret* many readers called to ask if I built and sold drywashers. I do not. However, in response to these inquiries, I decided to include construction plans in this book for a manually powered drywasher that incorporates many of the features of the drywashers I have built and used.

This drywasher uses a bellows that operates by a pull cord. It is as simple as they come. Another feature is a hopper to contain the paydirt. The hopper holds a full five gallon plastic bucket's contents. You must pre-screen, or classify, this paydirt before you feed it through the drywasher. I used a one-quarter inch square galvanized mesh that fit over the mouth of the bucket. The smaller the size of the particles of the paydirt, the better it is. Close sizing of particles improves recovery, but be very careful when screening gravel known to contain larger nuggets. Check the waste before you toss it away. The paydirt meters out through a slide door adjusted in the field for the material being processed. A clapper block and strike provide the jarring action to the paydirt that helps to settle the gold onto the muslin of the riffle tray.

The pull cord that operates the bellows may be of any length convenient to the operator. I use one four feet long and sometimes I hold it shorter so that I can watch the action of the gravel on the riffle tray. When I pull the cord, I limit the throw of the bellows, that is, the up and down distance the bellows travels. It is usually about one-half of the distance that the bellows can travel up and down. As I pull up the bellows I try to make the clapper block hit the strike a sharp blow. In operation, I soon establish a rhythm of about two pulls per second dependent on the property of the paydirt I am processing. It is important in operating a pull cord-operated drywasher to switch hands frequently to prevent fatigue. Operating a pull cord-operated drywasher is monotonous and boring; most manually powered drywashers are. The reward for enduring this is gold. My pull cord-operated drywasher that I used in the Las Guijas Mountains rewarded me with much, coarse gold including nuggets almost too large to pass through the screen. I was fortunate in that these larger nuggets passed through the diagonal of the one-quarter inch mesh. The gold that this machine recovered ranged from very small specks all the way up to those larger nuggets. Most of the gold was behind the upper three riffles. I felt that the machine recovered virtually all the gold in the paydirt.

When I go out to the gold placers with my drywasher I bring along one five-gallon container of water allocated only for panning the concentrate. I do not use my drinking water for panning. I fill a small plastic washbasin with water and use a small plastic finishing pan for panning. I do my panning only when testing. If I get good results, and if I'm there for several days, I hold off any panning until the end of each day. Panning takes up time that you should use for processing paydirt; you can always pan sometime later but you only have a limited amount of time for working the gravel.

When I'm on a trip of short duration, I test pan only to see if I'm into good color, then work diligently to accumulate as much concentrate as I can before I quit. Occasionally, as I dump out the concentrate from the riffle tray into my concentrate container, I see the gold on the riffles. I remove the larger nuggets and put these into my film container, but I do not pan out the concentrate in the field despite the temptation. I prefer, instead, to do this at my leisure back home. This may sound like severe discipline but it's not to me; everyone has his own way of mining. However, I do believe that it is important to take frequent rest, and water, breaks; and, I make it a point to have a relaxing lunch and enjoy my surroundings. Remember, we are here not only for the gold.

If you are handy with tools, building this drywasher is fairly easy. The only power tool that I use when building my drywashers is an electric drill. A variable speed drill of three-eighths inch size is adequate. I do all wood cutting with a back saw and miter box, or with a standard carpenter's saw, and sheet metal cutting with a pair of sheet metal shears. For bending sheet metal, I place it between two blocks of wood clamped tightly, and hammer it over. All the materials are locally available in most Arizona cities. I am assuming that if you decide to build this drywasher you will improve on it. Please, do.

DRYWASHER - CONSTRUCTION DRAWINGS

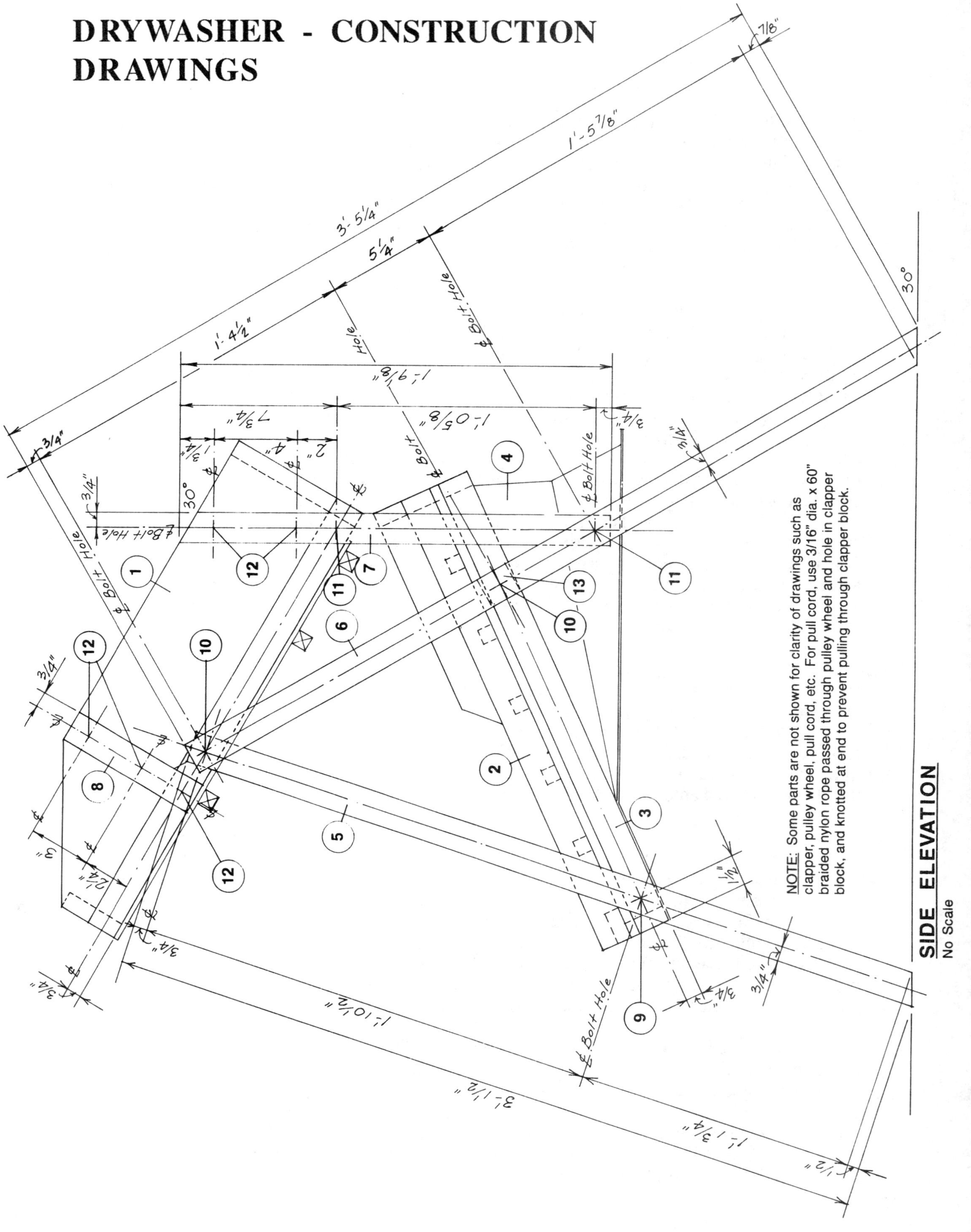

MATERIALS LIST FOR DRYWASHER ELEVATIONS

(1) - Hopper (see Hopper details)
(2) - Riffle Tray (see Riffle Tray details)
(3) - Bellow Box (see Bellow Box details)
(4) - Bellows, from 10 oz. canvas. Cut to fit as shown and glue and staple to inside of Bellow Box and to bottom of 1/4" plywood
(5) - 3/4" x 1-1/2" pine leg (each side)
(6) - 3/4" x 1-1/2" pine leg (each side)
(7) - 3/4" x 1-1/2" pine support brace (each side)
(8) - 3/4" x 1-1/2" pine support stake (each side)
(9) - 1/4" dia. x 2-1/4" carriage bolt with 1/4" dia. flat washer and 1/4" dia. lock nut
(10) - 1/4" dia. x 2-3/4" carriage bolt with 1/4" dia. flat washer and 1/4" dia. lock nut
(11) - 1/4" dia. x 2-1/4" hex bolt with 1/4" dia. flat washer washer at each end and 1/4" dia. lock nut
(12) - 1-1/4" long drywall screws
(13) - 3/4" x 1-1/2" x 1-1/2" spacer block nailed and glued to (6)

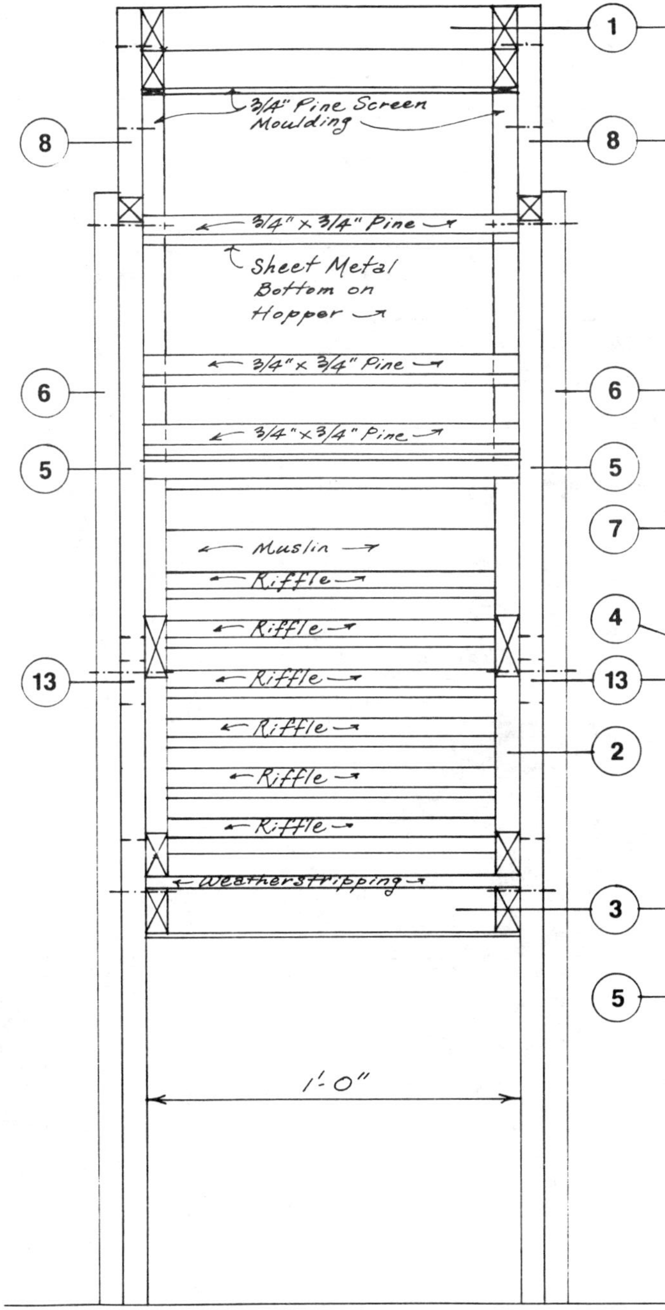

FRONT ELEVATION

No Scale

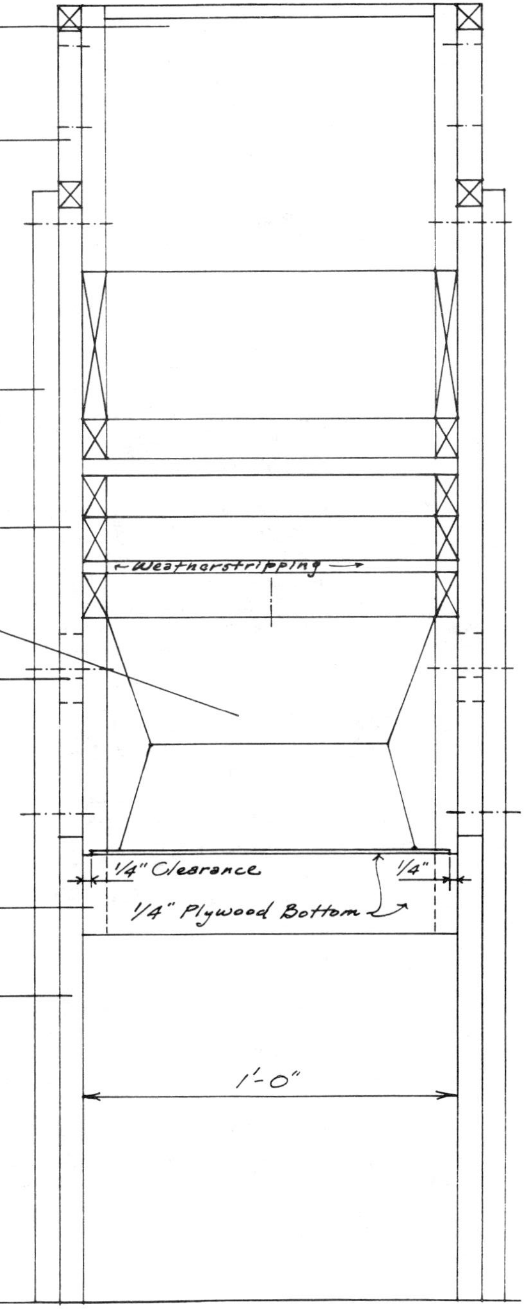

BACK ELEVATION

No scale

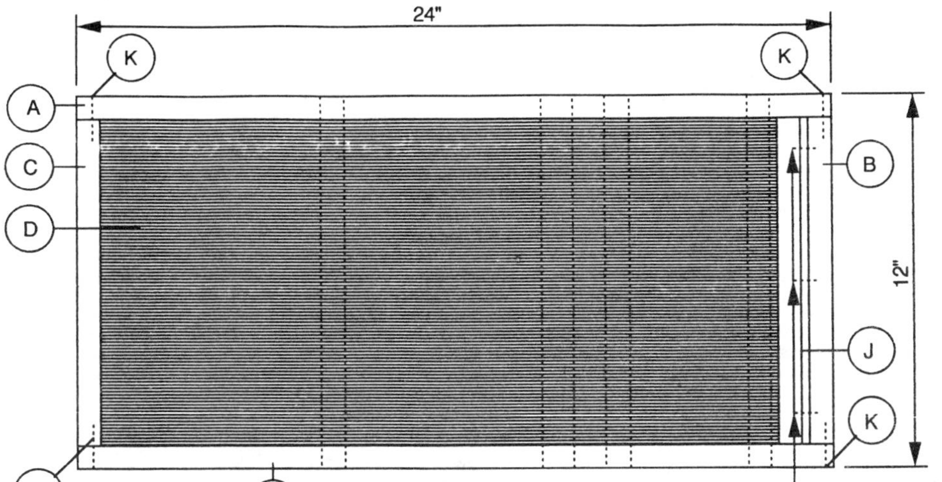

PLAN
3" = 1'-0"

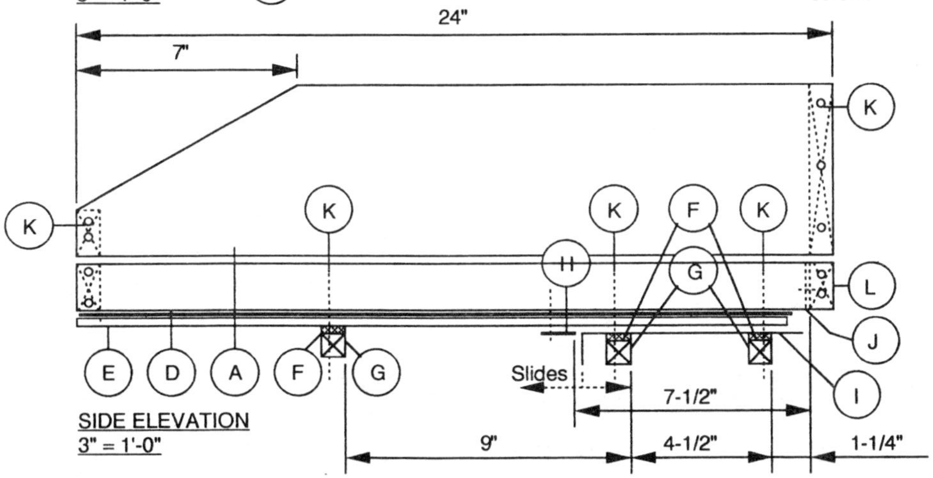

SIDE ELEVATION
3" = 1'-0"

NOTE:
Glue all joints with carpenter's glue

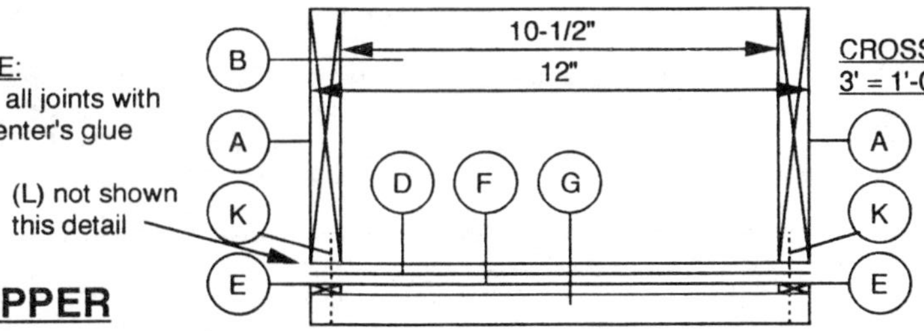

CROSS SECTION
3' = 1'-0"

HOPPER

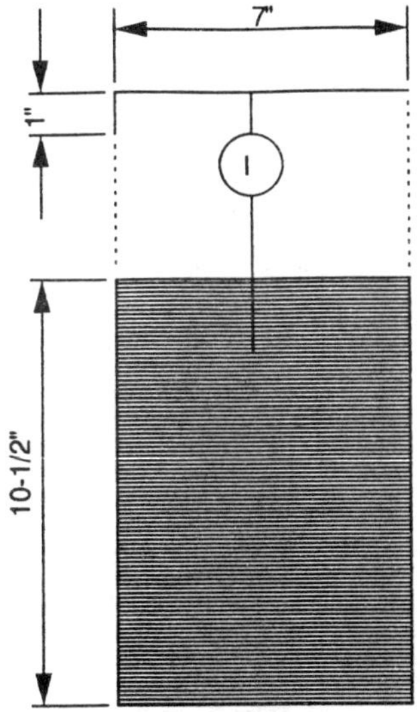

ELEVATION AND PLAN - SLIDE DOOR
3" = 1'-0"
(installs between (E) and snug against underside of (D) - slides freely)

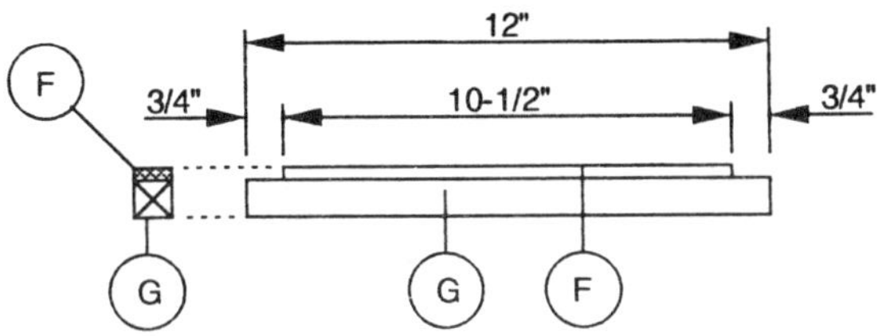

CROSSPIECE SUPPORT
3" = 1'-0"

MATERIALS LIST
(A) - 1" x 6" (nominal) x 24" pine
(B) - 1" x 6" (nominal) x 10-1/2" pine
(C) - 3/4" x 1-1/2" x 10-1/2" pine
(D) - 12" wide x 22-1/4" x 26 ga. galvanized sheet metal - nail to (L) w/brads spaced 6" on centers
(E) - 3/4" wide pine screen-moulding - use 3/4" flat head screws at 6" o.c. - counters'k
(F) - 1/2" wide felt weather stripping - glue to (G)
(G) - 3/4" x 3/4" x 12" pine crosspiece - screw to (L) with (K)
(H) - 1/8" x 1" x 12" flat aluminum - use sheet metal screw at each end to (L)
(I) - 26 ga. galvanized metal slide door
(J) - 26 ga. galvanized metal deflector
(K) - 1-1/2" long drywall screws
(L) - 12" x 24" frame from 3/4" x 1-1/2" pine

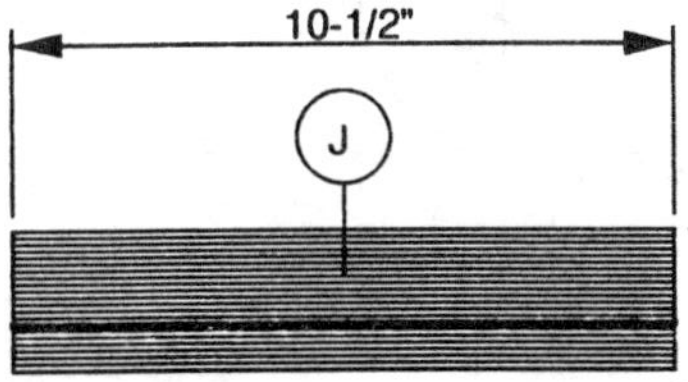

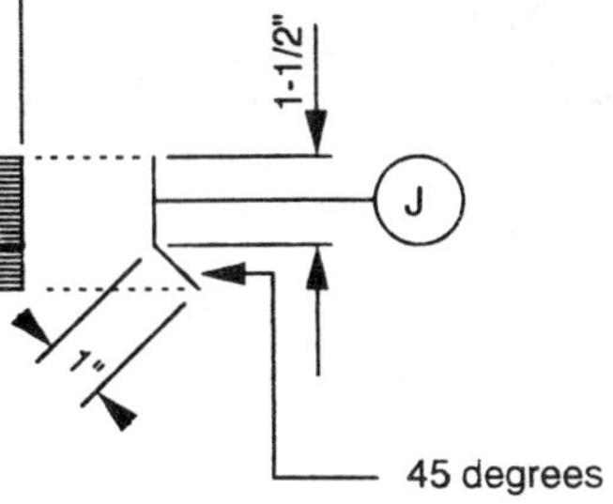

DEFLECTOR
3" = 1'-0"
(installs against (B))

HOPPER

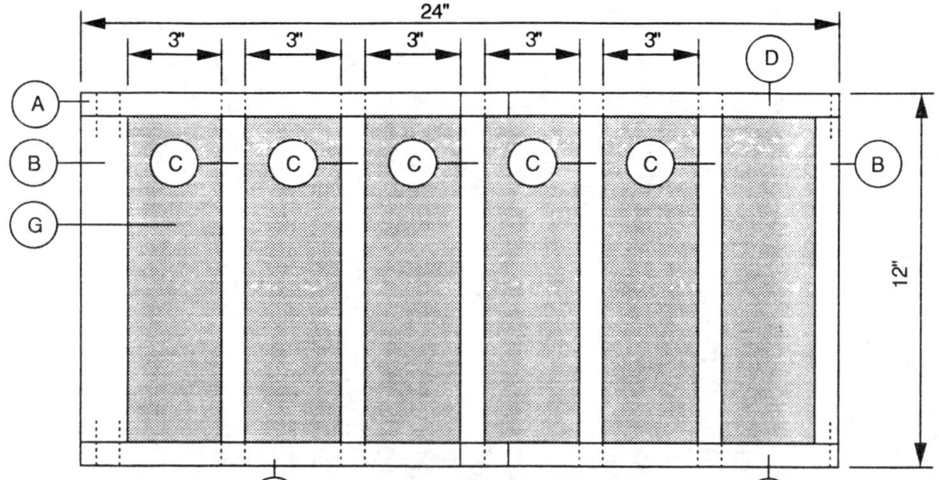

PLAN
3" = 1'-0"

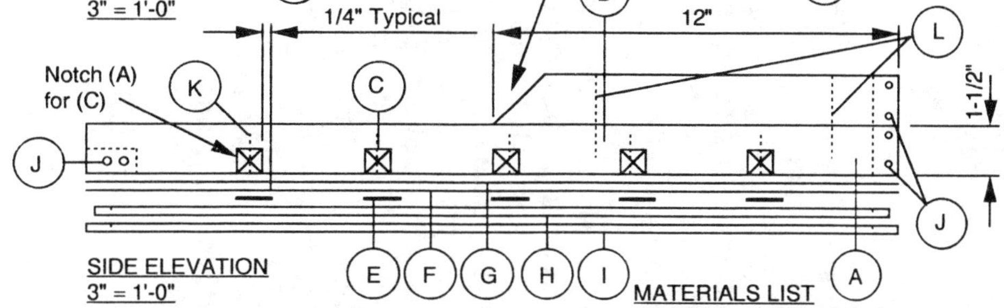

SIDE ELEVATION
3" = 1'-0"

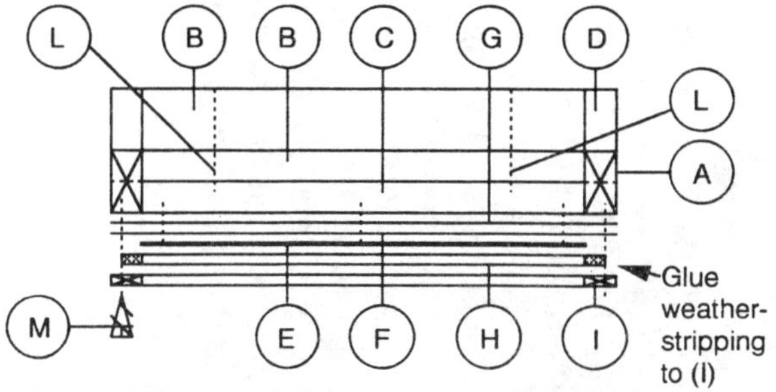

CROSS SECTION
3" = 1'-0"

NOTE:
Glue all joints with carpenter's glue

MATERIALS LIST
(A) - 3/4" x 1-1/2" x 24" pine
(B) - 3/4" x 1-1/2" x 10-1/2" pine
(C) - 3/4" x 3/4" x 12" pine
(D) - 3/4" x 1-1/2" pine (cut bevel @ 45 degrees)
(E) - 1/8" x 1" x 10-1/2" flat aluminum (screw to (B) with three 1/2" long FHWS, countersink heads flush)
(F) - 1/4" x 1/4" galv. hardware cloth
(G) - Muslin (staple to (A))
(H) - 1/2" wide felt weatherstripping
(I) - 3/4" wide pine screen-moulding
(J) - 1-1/2" long drywall screws
(K) - 1-1/4" long casing nails (glue & nail)
(L) - 2" long casing nails (glue & nail)
(M) - 3/4" long flat head wood screws @ 6" on centers, heads flush with (I)

RIFFLE TRAY

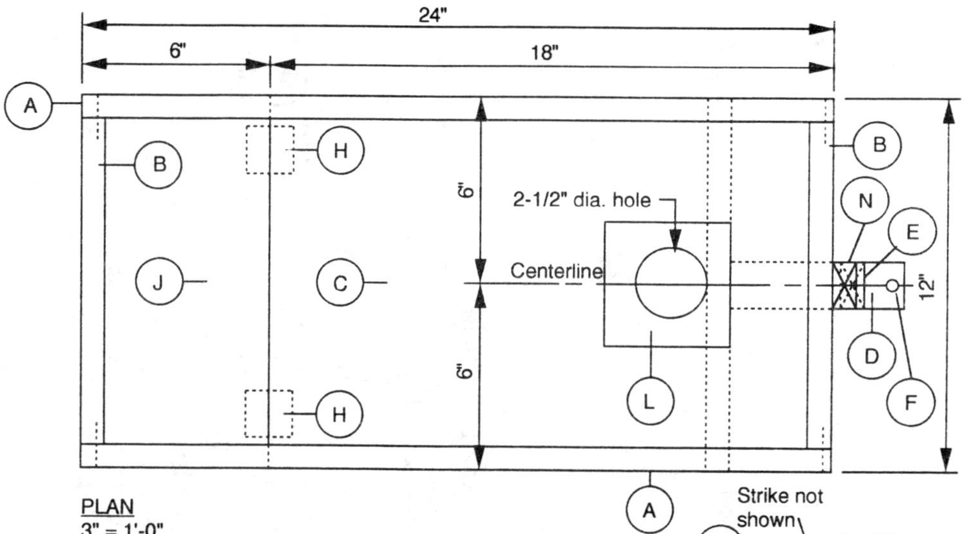

PLAN
3" = 1'-0"

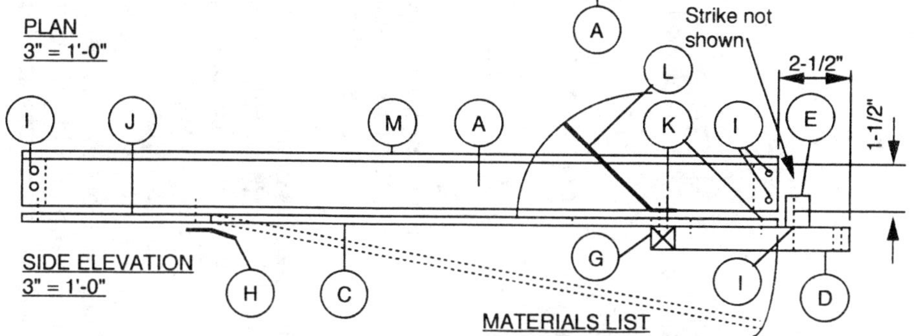

SIDE ELEVATION
3" = 1'-0"

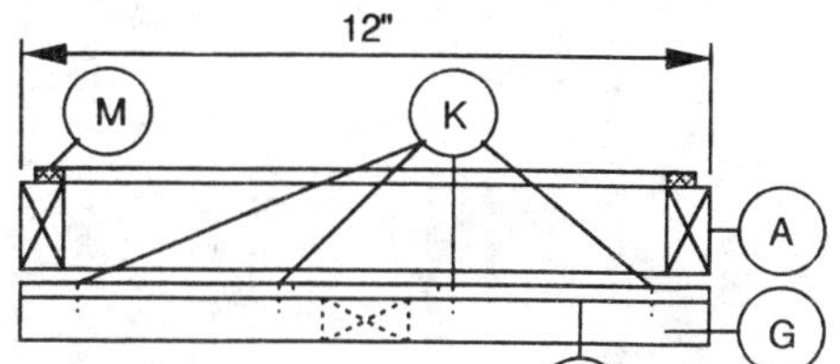

CROSS SECTION
3" = 1'-0"

NOTE:
Glue all joints with carpenter's glue

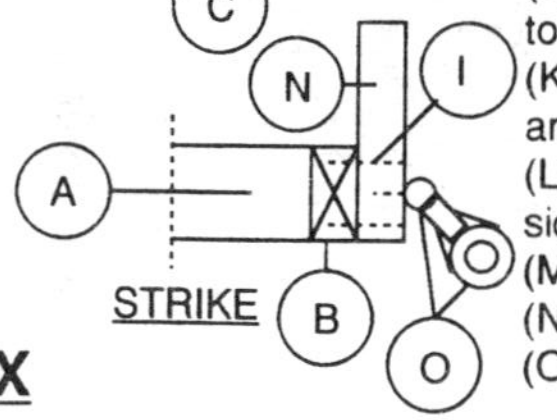

STRIKE

MATERIALS LIST
(A) - 3/4" x 1-1/2" x 24" pine
(B) - 3/4" x 1-1/2" x 10-1/2" pine
(C) - 1/4" A-D plywood (swings)
(D) - 3/4" x 1-1/2" x 5-1/4" pine, glued to (C)
(E) - 3/4" x 1-1/2" x 1" pine clapper
(F) - 3/16" dia. hole
(G) - 3/4" x 3/4" x 12" pine, glued to (C)
(H) - 1-1/2" x 1-1/2" cabinet hinge
(I) - 1-1/2" drywall screws (glue all joints)
(J) - 1/4" A-D plywood (fixed), nail and glue to (A) & (B)
(K) - 1/2" flat head wood screws through (C) and into (D) & (G), minimum 2 each piece
(L) - Leather flapper valve, stapled fixed side to (C)
(M) - 1/2" wide felt weatherstripping (glued)
(N) - 3/4" x 1-1/2" x 3-1/2" pine strike on c.l.
(O) - Roll-up-shade pulley wheel, screweye

BELLOW BOX

CHAPTER VIII

CLASSIFICATION - SCREENING MADE EASY

In any gold mining operation, whether large or small, it is necessary to classify the gravel for efficient gold recovery. Classification is just another name for screening the gravel to a predetermined size before processing it with the gold recovery equipment.

Since the size of the gold expected is considerably smaller than most of the gravel in which it occurs, it is extremely important to reduce the amount of barren material put through the gold recovery equipment. Classification is the method of reducing the amount of barren material to a workable minimum. A variety of devices accomplish this. There are grizzlies, constructed of a series of spaced parallel steel bars, or heavy steel mesh, and trommels, which are rotating steel barrels with holes up to three-quarters of an inch in diameter punched in the barrel. Other devices include punch plates of steel or aluminum usually used on suction dredge sluices, and ordinary screens made of stainless steel mesh, or galvanized hardware cloth. Openings range upwards to three-quarters of an inch, but are usually one-half-inch or smaller.

Ready-made classification screens that fit over the mouth of a five gallon plastic bucket are available from your local prospecting equipment dealer. These are strictly for the recreational prospector. Their only drawback is they are slower at classifying the gravel than the other devices. The reason for this is that they hold only a small amount of material for classification and they require manual shaking.

Although I still use classification screens fitting over plastic buckets, I wanted a screen capable of handling greater amounts of gravel, quicker, and with a minimum of hand labor. The excavation of gold-bearing gravel by manual methods constitutes the major item of work by the recreational prospector, and is the most time-consuming. Since time in the field is extremely precious, anything you can do to reduce labor, and conserve time, is to your benefit. The use of a portable grizzly screen will help you in this regard.

When I designed the portable grizzly screen shown on the following pages, I wanted it light in weight, portable, sturdy and efficient; and, I wanted to make it by myself out of relatively inexpensive materials using only hand tools. It had to take advantage of gravity to screen the material, and the screened material had to go directly into a container (a plastic bucket). It had to handle the gravel as fast as I could shovel it, with no further attention on my part. After many, many hours of use, it shows only a little wear and has met all of my expectations. I would not mine gold without it. The design of my grizzly screen uses a mesh of one-half-inch square. There are two considerations here. One is the size of nuggets expected and the other is the capability of the equipment. My equipment is capable of handling screenings of one-half-inch-minus size. In Arizona, the gold in most of the placer fields is small, with a few exceptions. Nuggets exceeding one-half-inch in diameter are rare; the majority of the gold is much smaller. In those locations where nuggets exceeding one-half-inch are quite possible, I go over all my screenings with a metal detector.

PORTABLE GRIZZLY SCREEN

MATERIALS LIST:

1. Six (6) - 3/4" x 1-1/2" x 8' Select Pine.
2. One (1) sheet 1/4" x 12" x 48" A-D fir plywood (cut from one sheet 24" x 48").
3. One (1) 12" wide x 48" long 1/2" x 1/2" mesh galvanized hardware cloth (cut from 24" wide roll).
4. Four (4) #12 screweyes.
5. Two (2) lengths of lightweight chain each 36" in length (cut from one 72" length) hooked onto screweyes each side of grizzly screen frame and legs.
6. Dozen 1-1/2" long drywall screws.
7. Half-pound 2-1/2" long galvanized casing nails or one-and-a-half dozen 2-1/2" long drywall screws.
8. Staples and small stapler for stapling mesh to frame.
9. Carpenter's glue.
10. Two (2) 1/4" dia. x 2" long hex bolts, with two flat washers (one each side of frame and leg). Tighten nut only to snug fit to permit set-up of legs.

Fig. 100. Prototype of author's Portable Grizzly Screen. Screened gravel slides down chute by gravity to plastic bucket set under semi-circular opening in chute. Screen accommodates buckets up to five gallon size. In practice, simply toss a shovel full of gravel onto the screen and let gravity do the rest.

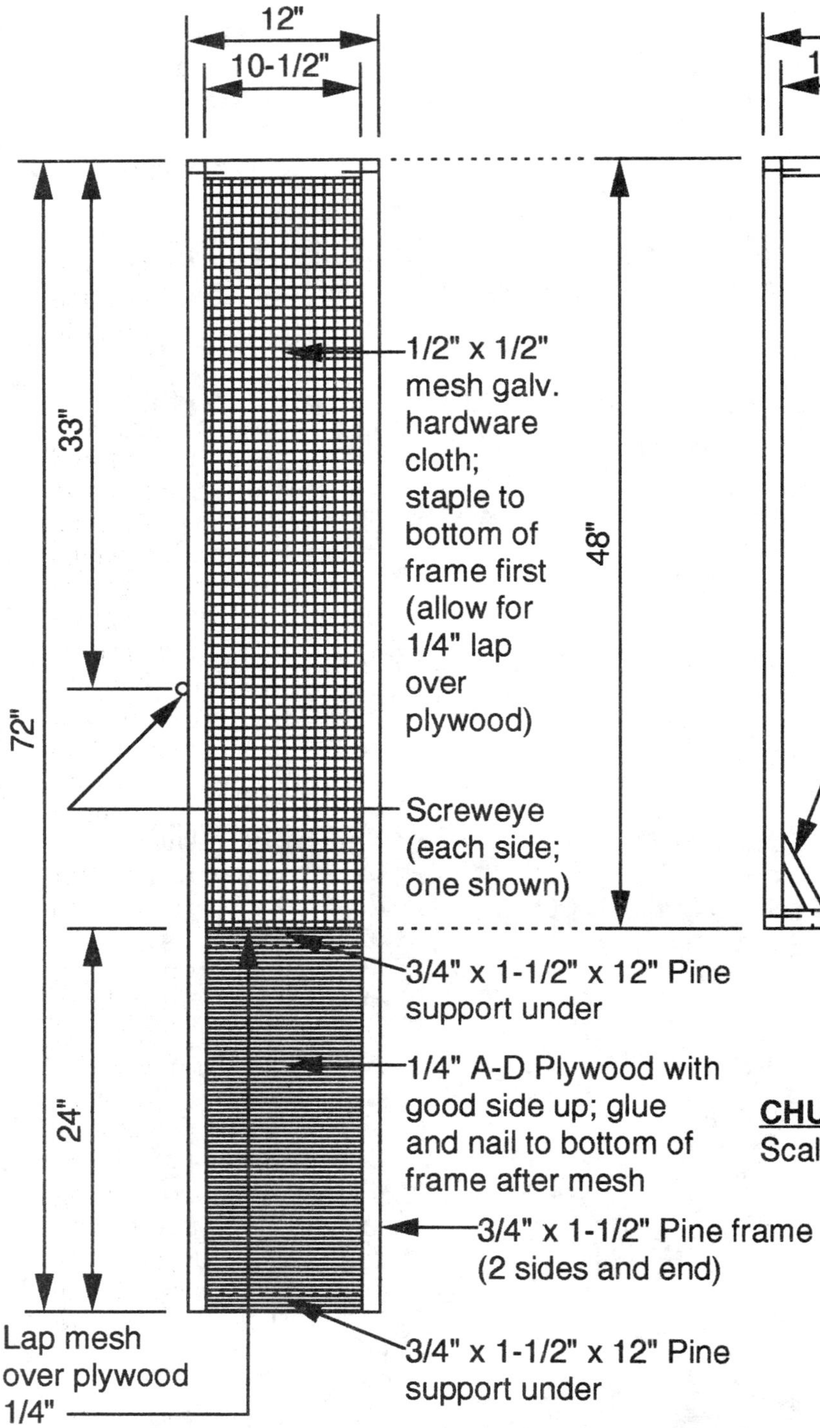

GRIZZLY SCREEN, AS SEEN FROM TOP
Scale: 1" = 1'-0"

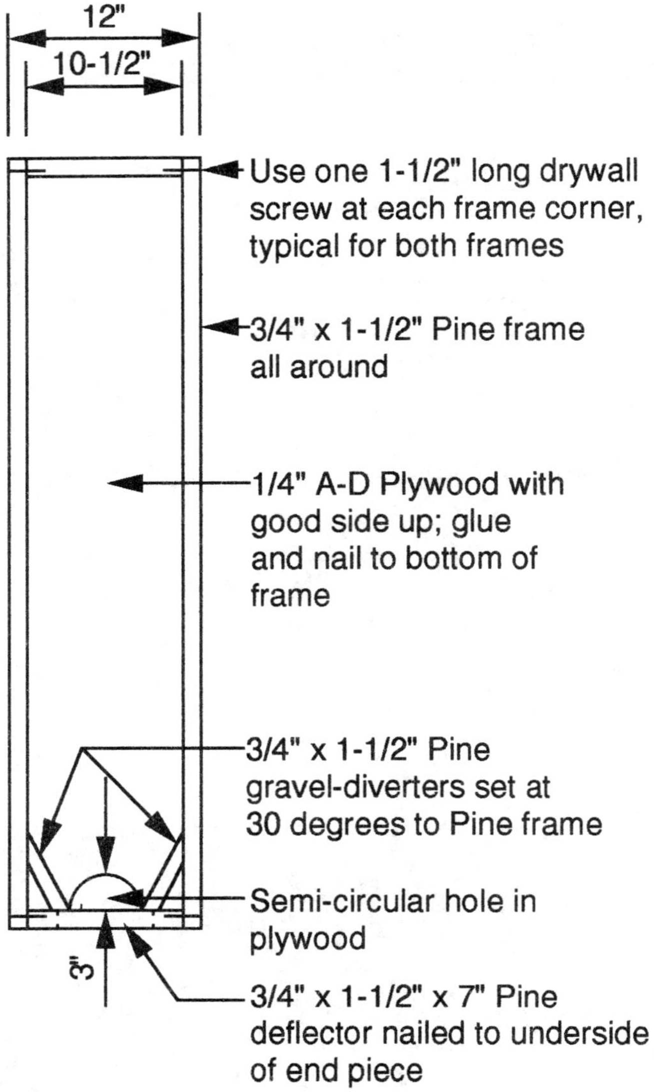

CHUTE, AS SEEN FROM TOP
Scale: 1" = 1'-0"

Note: Chute fastens to underside of grizzly screen frame. Use galvanized casing nails through pre-drilled holes in chute frame, or use 2-1/2" long drywall screws spaced approximately 8" apart.

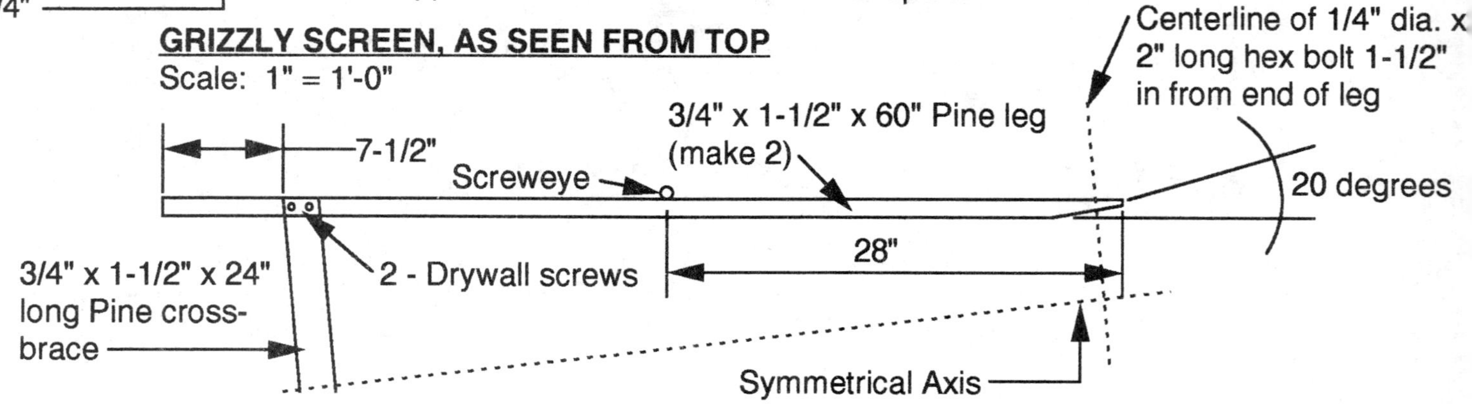

LEG AND CROSSBRACE
Scale: 1" = 1'-0"

KOFA NATIONAL WILDLIFE REFUGE

The Kofa National Wildlife Refuge was established in 1939 and is managed by the United States Fish and Wildlife Service. The refuge encompasses 660,000 acres of pristine desert environment which is the home of the unique desert bighorn sheep and the California palm, the only native palm in Arizona. Notable wildlife species found in the area include the white-winged dove, desert tortoise, and desert kit fox. The Kofa Mountain barberry (a rare plant found only in Southwest Arizona) occurs on the refuge.

Bighorn sheep are found chiefly in the two mountain ranges that dominate the landscape — the Kofa Mountains and the Castle Dome Mountains. Although these mountains are not especially high, they are extremely rugged and rise sharply from the surrounding desert plains, providing ideal bighorn sheep country. A wide variety of plant life is found throughout the refuge.

Water is always scarce in a desert. Natural water sources are highly variable and may not last until seasonal changes can replenish the supply. By enlarging natural water holes, shading them to reduce evaporation and blasting artificial basins in areas previously without a water supply, refuge managers have greatly increased the availability and reliability of water.

While all wildlife have benefited from these water developments, the desert bighorn sheep have especially responded to this assistance by producing a larger, healthier herd. Approximately 800–1,000 sheep now live in the refuge. In recent years, this prolific herd has provided animals for transplanting throughout Arizona and neighboring states.

Palm Canyon, in the west end on the Kofa Mountains, is well known for its native palms. These palm trees are probably remnants from a period of time when this area was wetter and cooler than at the present time. Though few in number, less than 100, this handful of trees plus a few scattered trees on the refuge and one or two other groves around the state are probably the only native palms in Arizona.

In the early part of this century, a number of mines were established in the mountainous areas of the refuge. One of the most notable was the King of Arizona Mine. It gave the Kofa Mountains their name — "Kofa" being contracted from "King of Arizona."

Most refuge roads are unmaintained trails and many of them are passable only to four-wheel drive vehicles. Other roads may be passable to two-wheel drive, high clearance vehicles (see map). Vehicles may be damaged by brush or rocks, or may become stuck in sandy areas. Before traveling on the refuge, visitors should contact refuge headquarters to learn of current conditions. There are no facilities on the refuge for towing, gasoline, sanitation, or drinking water.

Disposal of Waste

The dumping of litter, sewage, or liquid wastes on the refuge is prohibited.

Government Property

The destruction, injury, defacement, disturbance, or removal of any government property is prohibited.

Commercial Mineral Activity

The refuge is closed to mineral entry. Prospecting, removal, or disturbance of sand, rock, gravel, or minerals is prohibited. (Exceptions are valid mining claims existing prior to 1974.)

Disturbance of the Peace

Disturbance of the peace or other disorderly conduct on the refuge is prohibited. No person who is intoxicated shall enter or remain upon the refuge.

Pets

Pets are permitted only if they are confined. (Except the use of dogs when participating in a legal hunt.)

Treasure Hunting

Persons are prohibited from searching for or removing objects of antiquity, Indian artifacts, or paleontological objects.

Many outdoor recreational activities are permitted on the refuge with a few restrictions. The more popular pursuits are as follows:

Hiking, Sightseeing, Photography and Nature Observation

Permitted in all areas of the refuge except on patented mining claims and other private inholdings.

> CAUTION- Past mining activity has left numerous vertical shafts, drift tunnels and open pits throughout the refuge. These are extremely dangerous due to possible caving or collapsing. There is no practical way that they all can be located. Because of the extreme danger, no attempt should be made to enter or explore them at close range.
>
> Kofa was included in the massive desert military training exercises conducted by General Patton during World War II. Unexploded ordinance may be encountered during cross country hiking. DO NOT PICKUP anything that appears to be military hardware. Mark the location of the object and report it to refuge headquarters.

PUBLIC USE REGULATIONS

The following is a summary of refuge regulations taken from Title 50, Code of Federal Regulations. For additional detail or clarification, contact the refuge headquarters as below. All public access, use or recreational activity, unless specifically authorized, is prohibited.

Wilderness

Passage of the 1990 Arizona Desert Wilderness Act included 516,300 acres of the Kofa National Wildlife Refuge. Wilderness status acknowledges the relatively pristine character of the refuge and dictates some management restrictions. Visitors to the refuge may encounter wilderness boundary signs. No vehicular travel, including bicycles, is permitted into wilderness areas. Only foot or horseback travel is permitted.

Laws and Regulations

Any person may enter or use the Kofa National Wildlife Refuge in compliance with all public use regulations, the provisions of any special use permits, and official notices posted in the area. All Federal, State, and County laws, regulations, and ordinances are enforced on the refuge.

Vehicular Travel

Mechanized, vehicular traffic is limited to designated roads (see map on other side). Off road vehicle travel is prohibited. All vehicles, including ATC's, quadratrac and motorcycles and all operators must be licensed and insured for highway driving. Speed is limited to 25 MPH unless otherwise posted. Mountain bicycles are considered vehicles on the refuge.

Animal and Plant Life

Collecting, possessing, molesting, disturbing, injuring, destroying, removing or transporting any plant or animal or part thereof (alive or dead) is prohibited. (Except for legally taken game.)

Firearms and Explosives

Carrying, possessing, or discharging firearms or explosives (including fireworks) on the refuge is prohibited. Persons may use legal firearms for approved hunting provided that when they are being transported in vehicles they are unloaded and cased or dismantled.

Weapons Other than Firearms

The possession or use of crossbows, bow and arrows, air guns or other weapons is prohibited except when they are used for approved hunting.

Hunting

Limited hunting is permitted. Contact the Refuge Manager for additional information and current regulations.

Rockhounding and Rock Collecting

Recreational (noncommercial) rockhounding, including digging with simple hand tools, is permitted only in the designated area known as Crystal Hill.

On the remainder of the Kofa National Wildlife Refuge, rock or mineral collecting is restricted to materials that are exposed and collectable without the use of tools including metal detectors. Digging, including the use of simple hand tools, is prohibited.

Camping

Visitors may select their own campsites. However, applicable State law prohibiting camping within 1/4 mile of waterhole is in force. Vehicles must remain within 100 feet of designated roads. Backpackers may camp anywhere beyond the 100 feet limit keeping in mind the location of waterholes. All camping is limited to 14 days any 12 month period.

Fires

Campfires are permitted, but only dead wood may be used. Wood is scarce, please use it sparingly. No wood or tree stumps may be uprooted or removed from the refuge.

As a reminder, please contact the refuge headquarters for additional information and to report accidents or any unusual incidents or observations.

Kofa National Wildlife Refuge
356 West, First Street - P.O. Box 6290
Yuma, Arizona 85366-6290
Telephone: (602) 783-7861
Monday - Friday, 8:00 a.m. - 5:00 p.m.

APPENDIX

Refuge: A safe haven.

The Public Use Regulations at left apply to all persons entering upon the Kofa National Wildlife Refuge.

The purpose of the regulations governing permitted and prohibited activity on the refuge is to protect the native wildlife, particularly the desert bighorn sheep, and the pristine desert environment. Therefore, in the best interests of the native wildlife, you may not so much as dig a single grain of sand with a simple hand tool. *However*, in season, and with a proper license and permit, you may shoot, and kill, bighorn sheep, deer, quail, cottontail rabbit, coyote, bobcat, and fox! Apparently, digging with a simple hand tool, or even using a metal detector, is deemed more harmful to the wildlife than killing them. Hardly a refuge by definition. Yet, this is the case with most, if not all,wildlife refuges in Arizona.

Since recreational prospectors don't buy, and are not required to buy, licenses and permits as are hunters, they are prohibited from engaging in a rather benign activity while hunters are given carte blanche. The reason is plain, and simple: BIG MONEY! But, is it fair?
- author

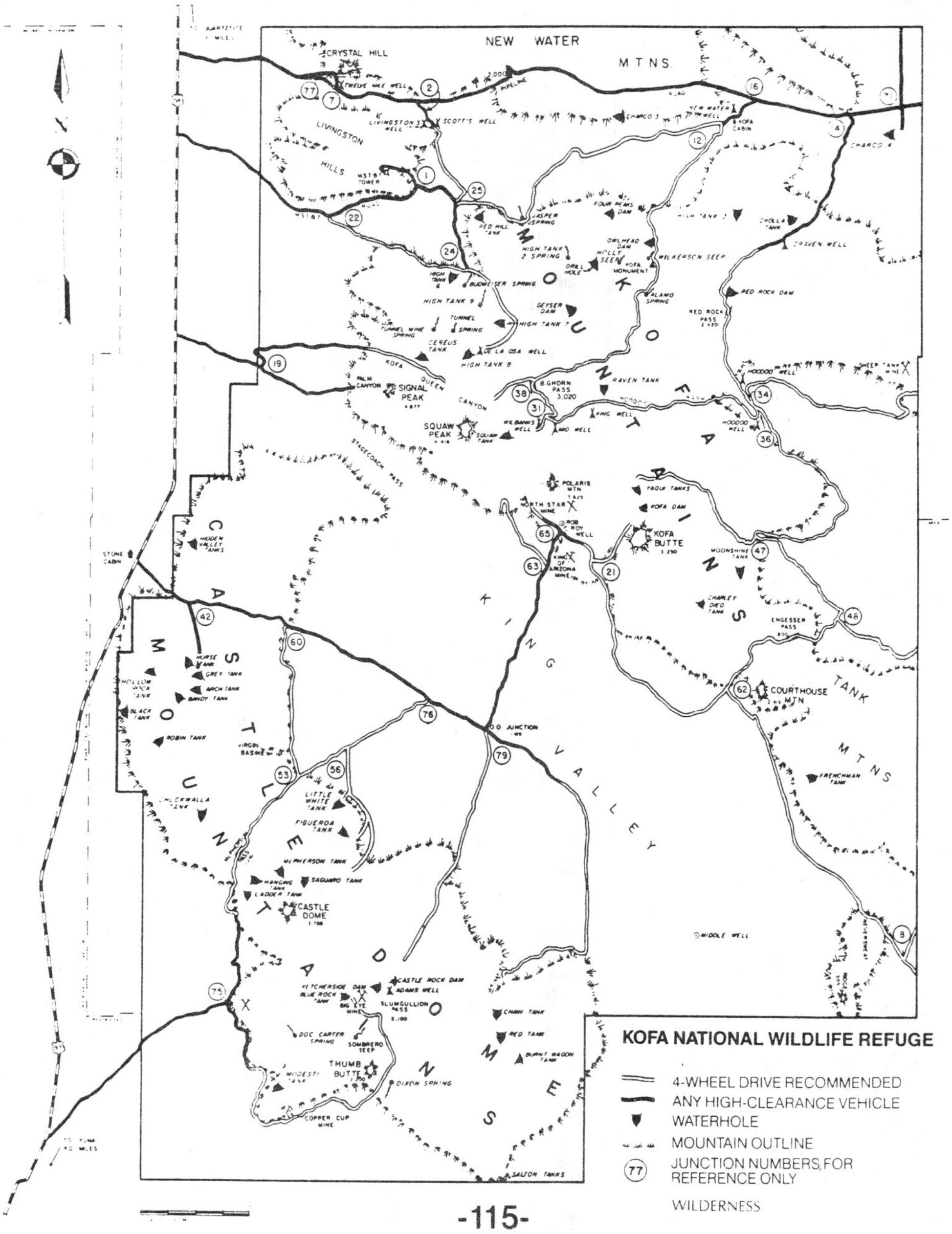
KOFA NATIONAL WILDLIFE REFUGE
4-WHEEL DRIVE RECOMMENDED
ANY HIGH-CLEARANCE VEHICLE
WATERHOLE
MOUNTAIN OUTLINE
JUNCTION NUMBERS, FOR REFERENCE ONLY
WILDERNESS
NEW WATER MTNS
CRYSTAL HILL
LIVINGSTON HILLS
KOFA MOUNTAINS
SIGNAL PEAK
SQUAW PEAK
POLARIS MTN
KOFA BUTTE
CASTLE MOUNTAINS
KING VALLEY
COURTHOUSE MTN
TANK MTNS
CASTLE DOME
CASTLE DOME MTNS
THUMB BUTTE